Creative Wheat Cookery

2

Creative Wheat Cookery

Over 300 Easy Tips, Tasty Recipes and Low-Cost Ideas for Using Wheat and Gluten in the Home

Evelyn C. Ethington

Copyright © 1975, 1999 By
HORIZON PUBLISHERS & DISTRIBUTORS, INC.

All rights reserved.
Reproduction in whole or any parts thereof in any form
or by any media without written permission is prohibited.

Fourth Printing: February, 1999

International Standard Book Number
0-88290-658-5

Library of Congress Catalog Card Number
75-5321

Horizon Publishers' Catalog and Order Number
1242

Printed and distributed
in the United States of America by

Mailing Address:
P.O. Box 490
Bountiful, Utah 84011-0490

Street Address:
50 South 500 West
Bountiful, Utah 84010

Local Phone: (801) 295-9451
Toll Free: 1 (866) 818-6277
FAX: (801) 295-0196

E-mail: horizonp@burgoyne.com
Internet: http://www.horizonpublishers.biz

Contents

Acknowledgments vii
About The Author viii
1. The Ingredients Of Good Nutrition 9
   Basic Facts About Food 9
   Six Essential Nutrients 10
2. Wheat: The Basic Food 13
   Early History 13
   Present Production and Use 13
   Classification and Description 14
   What Is Gluten? 14
   What Is Triticale? 15
   Types of Protein 15
   Wheat Is Economical and Nutritious 16
3. Grinding and Storing Wheat in the Home 18
   Grinding and Milling Methods 18
   Tips for Choosing a Grinder 18
   Choosing The Wheat You Buy 19
   Storing and Preserving Wheat 19
4. Cooking Helps 21
   Abbreviations 21
   Metric Conversions of Kitchen Measurements 21
   Oven Cooking Temperatures 22
   Adjusting for Altitude Differences 22
   Ingredient Substitutes 23

Section One
Using Wheat

5. Cooking Whole, Cracked And Bulgar Wheat 26
   Whole Wheat 26
   Three Methods For Cooking Whole Wheat 26
   Two Tasty Ideas For Using Cooked Whole Wheat 26
   Wheat Kernel Casserole 26
   Hot Wheat Drink 27
   Cracked Wheat 27
   Cooked Cracked Wheat 27

Cracked Wheat Casserole	27
Bulgar Wheat	28
Preparing Bulgar Wheat	28
Bulgar Cereal	28

6. Sprouting Wheat ... 29
- Sprouting Directions ... 29
- Sprouted Wheat Cereal or Treat ... 29
- Wheat Grass ... 30

7. Bread Making With Whole Wheat Flour ... 31
- Basic Bread Ingredients ... 31
- Kneading or Pounding Dough ... 32
- Pounding Method ... 33
- Forming Loaves in Bread Pans ... 34
- Causes of Poor Quality Bread ... 34
- Tips For Using Old Bread ... 35
- Basic Recipe ... 35
- Versatile Whole Wheat Yeast Bread ... 35
- Rye Bread ... 36
- Sliced, Buttered Bread Ring ... 36
- Crispy Buttered Rolls ... 36
- Stollen Bread ... 36
- Stick Bread ... 37
- Dill Bread ... 37
- Lemon Nut Bread ... 37
- Donuts or Scones ... 37
- Crunch Top Rolls ... 38
- Breakfast Cake ... 38
- Sweet Rolls ... 38
- Cinnamon Bread ... 39
- Whole Wheat French Bread ... 40
- Whole Wheat Rolls or Buns ... 40
- Hot Dog Buns ... 40
- Fancy Rolls ... 41
- Basic Recipe ... 41
- Basic Wheat Mix [BWM] for Quick Bread ... 41
- Biscuits ... 41
- Pancakes ... 42
- Waffles ... 42
- Dumplings ... 42

Section Two
Using Gluten

8. Gluten: Easy to Make, Easy to Use **45**
   Advantages of Cooking with Gluten 45
   Calculating Gluten Quantities 46
   Extracting Gluten in Three Easy Steps 47
   Baking and Grinding Gluten for Quick Use 48

9. Gluten in Place of Meat **49**
   Ground Gluten as a Substitute for Hamburger 49
   Best Ever Lasagna 49
   O-So-Yummy Puffs 50
   Favorite Winter Chili Pot 50
   Versatile Meatless Balls or Patties 50
   Wheat Franks 51
   Wheat Balls 51
   Tasty Baked Beans 52
   Beans from Scratch with a Vegetable Relish 52
   Quick Spaghetti Casserole 52
   Spanish Cornbread 53
   Sausage Casserole 53
   Quick Potato Casserole 53
   Oriental Casserole 54
   Sauce for Spaghetti 54
   Delicious Meatless Enchiladas 55
   Tempting Wheat Salad 55
   Nutritious Pet Food 56

10. Gluten as Part Meat **57**
   Gluten as a Meat Extender 57
   Recipes 57
   Versatile Meat Balls or Patties 57
   Special Franks 58
   Franks in Spicy Rice 58
   Pork Sausage Extender 58
   Meat Loaf With Applesauce 58
   Sausage Skillet Dish 59
   Hawaii Chicken With Meat Balls 59
   2-Tasty Meatloaves in One Mix 59
   Meat Loaf With A Special Sauce 60
   Meat Loaf and Gravy 60

Ground Round Soup 60
Quick Meat Pie 61
Enchiladas Casserole 61
Easy Skillet Chili 61
Well-Seasoned Noodle Casserole 62
Super Supper In-A-Dish 62
Quick Luncheon Dish 62
Ground Round Special 63
Chicken Layered Casserole 63
Turkey or Chicken Salad 63
Tuna-Chicken Bake 64
Quick Make-Ahead Tuna Casserole 64
New Way Tuna Souffles 64
Spanish Dinner 65
Meal-In-One Salad 65
Pizza Burger 65
Sour Cherry Pilaf 66
Iranian Koresh 66
Canned Tomatoes Layered Casserole 66
Barbecue Hamburger 67
Chili Balls 67
Turkey or Chicken Stuffing 67
Chicken Bake Supreme 68

11. Other Gluten Recipes 69
Mock Steaks 69
Chicken Fried Steak 69
Breaded Veal Steaks 70
Great Ideas for Lunchmeat 70
Spicy Lunchmeat 70
Sandwich Spread 71
Cook a Little Magic with Raw Gluten 71
Mini Pizza 71
Mock Swiss Steak With Vegetables 72
Treats for Snacking 72
Tasty Wheat Roast 72
Jerky 73
Miracle Wheat Jerky 73

Section Three
Using Bran

12. Using Bran and Starch Water 77
    Bran is Beneficial 77
    Cereal Flakes 77
    Caraway Crackers 77
    Quick Pancakes 78
    Whole Wheat Yeast Muffins 78
    Company Brownies 78
    Fresh Apple Pudding 78
    Brown Sugar Squares 79
    Chewy Bars 79
    Delicious Fudge Nut Bars 79
    Banana-Bran Bread 80
    Good Whole Wheat Danish Puffs 80
    Raised Whole Wheat Pizza Dough 81
    Bran Batter for Wheat Balls 81
    The Crown of Cornbread 81
    Cream Puffs 82
    Canned Old Fruit Cake 82
    Starch Water 82
    Be Thrifty: Use Starch Water 82
    Starch for Thickening 83
    Good Brown Gravy 83
    Handy Crackers 83
    Corn Tortilla 83

Section Four
Using Sauces & Toppings

13. Sauces 86
    Special Sauces 86
    Hawaiian Sweet and Sour Sauce 86
    Quick Sauce 86
    Curry Sauce 86
    Mushroom Sauce 87
    Bacon-Mushroom Sauce 87
    Rich Flavor Sauce 87
    Mustard Sauce 88
    Easy Mild Sauce 88

Tomato Soup Sauce 88
Cranberry Sauce 89
14. Toppings 90
Tasty Toppings 90
Seasoned Toppings 90
Seasoned Topping Number 1 90
Seasoned Topping Number 2 90
Seasoned Topping Number 3 91
Recipes For Using Seasoned Toppings 91
Tuna Salad 91
Mayonnaise Dressing 91
Sea-Food Dressing 92
Creamed Eggs 92
Stuffed Eggs 92
Dip For Raw Vegetables 92
Casserole Toppings 93
Blender Thousand Island Dressing 93
Quality Blue Cheese Buttermilk Dressing 93
Sweet Toppings 94
Cocoa Topping 94
Cinnamon Topping 94
Brown Sugar Topping 94
Recipes for Using Sweet Toppings 94
Double Quick Pie Crust 94
Frozen Pie 95
Fruit Pie 95
Simple Ice Cream Pie 95
Frozen Fruit Pie 95
Pineapple Dessert 96
Slice Date Cookies 96
Special Fruit Salad 96
Cream Cheese Cake 97
Marshmallows 97
Refreshing Peppermint Dream 97
Ready-To-Go Cereal 98
Marshmallow Squares 98
Superb Cookies or Pie Crust 98
Creamy Vanilla Ice Cream 98
Family Home Evening Treat 99

ACKNOWLEDGMENTS

Grateful acknowledgment is made to loved ones and friends for their valuable ideas and recipes.

My deep appreciation is extended to Mrs. Alma P. Burton and Mrs. Royce P. Flandro. But for their interest and persistent encouragement this book might not have been written.

To my favorite critics, my loving husband, Martin, and sons Layne, Brent and Paul, I give special thanks for their much help and love.

ABOUT THE AUTHOR
EVELYN C. ETHINGTON

Through extensive research, experimentation and cooking with wheat, Evelyn Ethington has become an expert in creative wheat cookery. She has given numerous lectures and demonstrations on the varied uses of wheat in her home, to club and church groups, and in the arts & crafts department of Education Week at Brigham Young University in Provo, Utah. She's developed many of the recipes in this book, particularly those using gluten. Because of the high cost of protein foods and the shortage of this necessary nutrient in today's diets, she has conducted extensive research and recipe experimentation with gluten as a protein source.

Mrs. Ethington is a native of a rural Louisiana community and the second of a family of ten children. Her background as the daughter of a farmer and cattle rancher was the source of her early interest in this field.

Her husband is retired from twenty-two years of active duty in the Air Force and is employed at Brigham Young University. They have lived in many states and are now settled in Orem, Utah. Mr. and Mrs. Ethington are the parents of four children, and have five grandchildren.

Chapter 1

THE INGREDIENTS OF GOOD NUTRITION

Do you consider yourself well fed? Are you providing your body with the right kinds of food so that it will function at its maximum efficiency? Many people today have enough to eat but are still not well nourished. They have high calorie diets with low nutritional value. A daily diet of nutritious foods can make the difference in how you look and feel. A conscientious effort to supply your body with proper nutrients can be the determining factor between good health and poor health. You don't have to be a professional dietitian to have the basic facts about your body's nutritional needs and how to supply them.

Basic Facts About Food

Let's look at some of the key ideas about food which every homemaker should understand:

1. Your body is a machine and, as such, needs energy to function properly. This energy comes from eating enough of the right food.

2. The amount of energy produced in the human body by the food you eat is measured by a unit of heat called a calorie. Your caloric needs depend largely on your sex, activity, age, body build and thyroid level.

3. The process by which your body converts food and water into energy and living tissue is called metabolism.

4. Foods are grouped according to the nutrients they provide. The following four are the basic food groups usually recommended by food scientists:

GROUP 1–DAIRY FOODS

Daily Individual Requirement:

    One to two cups or equivalent in milk products.
    (More is suggested for children.)

GROUP 2- MEAT, FISH, POULTRY AND EGGS

Daily Individual Requirement:

Two or more servings a day.

GROUP 3- VEGETABLES AND FRUITS

Daily Individual Requirement:

Four to five servings a day.

GROUP 4- BREADS AND CEREALS

Daily Individual Requirement:

Three to four servings a day.

Six Essential Nutrients

These basic food groups contain six essential nutrients which help to nourish, build and repair your body. These nutrients are:

1. CARBOHYDRATES—

Carbohydrates furnish calories which provide energy for our bodies to breathe, work and move about. Carbohydrates consist of starch and sugar and are found in such foods as potatoes, rice, wheat, grains, honey, fruits and vegetables. Common sources of starch are potatoes, rice and wheat. Sugar is found in honey, fruits and vegetables. If you do not eat enough carbohydrates, the fats and proteins you eat will be changed into glucose for the energy your body needs. If you eat too many carbohydrates, they will be converted to fat.

2. FATS—

Fats contribute more than twice the calories of carbohydrates. They are important for the protection and insulation of your body. They are found in all meats, oils, butter, margarine, shortening, whole milk, cheese, plant seeds, vegetable oils, nuts, olives and fish.

3. PROTEINS—

Proteins are most important for the growth, repair and replacement of cells for your body. They are also the source of nitrogen, carbon, hydrogen and oxygen to your body. A protein is made up of a chain of substances called amino acids. Some amino acids can be manufactured by the cells of the body but others must be supplied by protein foods. Amino acids cannot be stored in the cells of the body, so each day your diet must include sufficient requirements of protein foods. Foods rich in protein are beef, poultry, fish, milk and milk products, wheat and other grains, soybeans and other beans and nuts. The following chart gives recommended daily needs of protein for various ages. These figures are based on half the protein being animal protein.

Children	Ages 1-10	need 23-26 grams protein.
Boys	Ages 10-18	need 36-54 grams protein.
Men	Ages 18-75	need 60-65 grams protein.
Girls	Ages 10-18	need 36-48 grams protein.
Women	Ages 18-75	need 48-50 grams protein.
Pregnant		need +30 grams protein.
Lactating		need +20 grams protein.

4. MINERALS—

Minerals (salts) help perform many essential functions in the body and are needed for general good health and well-being. The common minerals needed are calcium, phosphorus, sodium potassium, copper, zinc magnesium, chlorine, sulfur, iron, cobalt, manganese and iodine. Foods rich in minerals are milk, wheat and whole grains, meats, fish, green vegetables, liver, raisins, iodized salt and seafood.

5. VITAMINS—

Vitamins keep the body functioning properly and are found in small amounts in all foods. There are many categories of vitamins and each performs a certain function for the body. Usually a well-balanced diet with varied foods will contain all the vitamins needed.

6. WATER—

Water is also considered a nutrient and is a necessary part of all living tissue. Water assists in the digestion of food, distribution of heat in the body, transporting needed substances to different parts of the body and in exchanging oxygen and carbon dioxide.

With the application of this basic health knowledge, you can strive for better health. The degree of good health you enjoy depends largely upon the quality and quantity of the foods you choose to eat

Chapter 2

WHEAT: THE BASIC FOOD

Early History

One of the oldest natural foods known to man is wheat. It has been linked to the development, progress and prosperity of many countries of the world. Although its origin is not really known, many scientists think it was first grown in Asia where the country of Iraq is now found. The Bible records early uses of wheat. Around 300 B.C. a Greek philosopher mentioned wheat that was grown around the Mediterranean Sea in his writings. In 1493 Columbus brought wheat to the Western Hemisphere.

Present Production and Use

Because of its nutritional value, economy and versatility, wheat has become the most important grain grown in the world. Wheat and wheat products are the main food in the diet of millions of people. In our country, whole wheat grain and enriched wheat products provide one-fourth to one-third of the protein, thiamine, iron, niacin and riboflavin in the average diet. In the United States, it is estimated each person consumes about 118 pounds annually. Farmers produce around 9,000,000,000 bushels annually which is more than the production of rice.

The U.S. Department of Agriculture lists 10 leading wheat growing countries in this order: Russia, United States, China (mainland), Canada, France, India, Italy, Turkey, Australia, Argentina. Of the more than 30,000 species of wheat grown in the world about 200 are grown in our country. The U.S. Department of Agriculture has collected approximately 15,000 varieties.

An enormous amount of research has and is still being done to develop uses for this grain. Today we enjoy a variety of breakfast food in ready-to-eat and cooked cereals. Macaroni, spaghetti, noodles, breads, rolls, crackers, and pastry are all made from wheat. Wheat germ and wheat oil are extracted from the embryo. Many delicious protein dishes can be made from wheat gluten as a substitute for other expensive protein products.

The whole wheat grain and enriched wheat products provide one-fourth to one-third of the protein, thiamine, iron, niacin, and riboflavin in the average American diet.

Classification and Description

Wheat belongs to the grass family Graminese and is classified in the cereal group. It will grow in hot, cold and dry climates. The slender, long, leaf-growing plant is green and ripens to a golden brown. The kernels are contained in the wheat head at the very tip of the main stem. A husk forms around the kernel with hair-like beards. One single plant yields about 50 kernels. The kernel, about one-fourth inch long, is divided into three parts: (1) germ- the embryo and life of the plant which is located at the tip end of the kernel; (2) bran- the outer layer; and (3) endosperm- the center. These three parts contain the following minerals and vitamins:

Minerals		*Vitamins*
Calcium	Iodine	Thiamine B-1
Magnesium	Sodium	Niacin
Iron	Fluorine	Riboflavin B-2 or G
Potassium	Silicon	Pyridoxine B-6
Phosphorus	Chlorine	Pantothenic Acid
Manganese	Boron	Biotin or H
Copper	Barium	Inositol
Sulphur	Silver	Folic Acid
And Other Trace Minerals		Choline
		Vitamin E

What Is Gluten?

Gluten is the protein of wheat and can be used as a good source for nutritious meals. One cup of raw gluten contains seventy-two grams of protein. This is much more than other common foods. For example:

1 egg contains 6 grams of protein.

1 oz. hamburger contains 7 grams of protein.

1 c. liquid milk contains 9 grams of protein.

1 oz. Process American Cheese contains 7 grams of protein.

4 tbsp. textured vegetable protein (T.V.P.), dry contains 10 grams of protein.

Gluten is best used in combination with other protein foods, however, so the body receives a complete supply of the various types of protein. A simple rule to remember in using gluten is to let 10 to 25 percent of the total bulk of the dish be other protein foods such as milk, cheese, eggs, meat, chicken, tuna or legumes. Most of the recipes in this book using gluten as a main dish have followed this simple rule.

What Is Triticale?

Triticale is a new grain developed by scientists. It is a hybrid—a cross between wheat and rye. Its kernel is about one-third larger than the average wheat kernel. Like wheat, it has a nutty-like flavor. The best varieties of Triticale have a protein content which varies between 13% and 18%, but its gluten content is low, making it necessary to mix it with wheat flour for making bread.

Because of its extensive root system, Triticale has the ability to reach more soil minerals. This gives it a higher mineral content than other cereal grains; as a result its price is about 25% higher than wheat.

Types of Protein

The main types of protein found in plants are albumin, globulin, glutelin, prolamine protamines and histones. Wheat contains high contents of glutelins and prolamines.

A distribution of the four main types of protein found in wheat is given as the percent of total seed protein represented by the individual protein types:

Albumin 3-5	Glutelin 30-40
Globulin 6-10	Prolamine 40-50

The glutelins and prolamines are listed as having lower contents of some amino acids, namely lysine, histidine and arginine. For the body to receive a balanced protein diet, it is necessary that wheat dishes be supplement with other protein food such as meat, fish, poultry, milk, cheese, eggs and legumes.

The Kjeldahl procedure has become the standard method for determining the protein nitrogen of grains. Through this procedure, chemists have been able to determine the protein content of wheat flour by mixing sulfuric acid and potassium with a sample of whole wheat flour. With the application of heat the nitrogen is isolated and then measured.

Amino Acid Compositions

Amino acids are the chief components of proteins. Chemists have measured the amino acid content in the various types of wheat. The following is a list of amino-acid compositions of certain wheat varieties. Data is expressed as grams of nitrogen in individual amino acids.

Amino-Acids	*Gluten (Wheat Kernel)*
Amide-N	No Determination
Aspartic Acid	1.8
Glutamic Acid	17.0
Proline	8.1
Glysine	3.3
Alanine	1.7
Valine	2.6
Leucine	3.8
Isoleucine	2.4
Phenylalanine	2.3
Tyrosine	1.6
Tryptophan	0.8
Serine	3.3
Threonine	1.5
Cysteine	1.6
Methionine	0.8
Lysine	1.7
Histidine	3.2
Arginine	6.1

Wheat Is Economical and Nutritious

Using whole wheat grain in your diet brings major benefits, both nutritionally and financially.

There are 26 known minerals and vitamins in the whole wheat grain which can add to your family's health and vitality. Many common health problems are linked to a lack of whole grains in the diet. Nutritionists are advising that we eat more whole wheat bread, saying that the roughage provided by the whole wheat kernel aids in proper elimination of body wastes.

One pound of wheat contains 1600 calories which can be considered ones daily energy requirement. A bushel of wheat will provide food for about 45 days for one person. There are approximately 1200 one-half cup servings of cracked wheat in 100 pounds of wheat. Even at today's inflated prices, wheat is the best buy on the market for daily use and for emergency food storage.

During this time when food prices are rising so rapidly, grinding wheat and other dry grains will provide substantial financial savings. Stone ground whole wheat bread can be made for about one-fourth the cost of store-bought bread. Cracked wheat cereal is about one-tenth the price of prepared cereal. Your investment will be returned to you time and time again over the years.

Chapter 3

GRINDING AND STORING WHEAT IN THE HOME

The most economical way to purchase wheat is to buy it in bulk in large quantities and to store it in the home. The most effective way to prepare the wheat for cooking, while preserving both its nutritional and economic value, is to process it with a wheat grinding mill. The purchase of an electric wheat grinder will open up a whole new field of wholesome cooking.

Grinding and Milling Methods

Let's take a look at wheat as it is ground. The wheat germ makes up about two percent of the wheat kernel. It contains much of the flavor, proteins, vitamins and minerals. The stone grinding method seems to have its advantage in that the wheat kernels are rubbed between two thick stones. This method spreads the nutritious wheat germ evenly in the flour.

Another method, which uses the roller, hammer and steel buhr mill, causes a break of the wheat kernel. This separates the germ, which remains an oily substance.

Most commercial flours have been de-germed to prolong shelf life. Most white flours contain preservatives, whiteners, agers, softeners, fresheners and spoilage preventers. With a home-owned stone ground mill you can be assured none of these are added, and that you are receiving all the benefits from the whole kernel, not just a part.

Other grains such as corn, rice, oats, millet, barley and rye that are dry and not too oily may be ground in a wheat mill, too. Legumes may be ground if they are completely dry. To be sure they are dry, place them in an oven at 200 degrees for about one hour. Grind first with stones adjusted for cracked wheat, then move the stones closer for finer grinding.

Many recipes call for a combination of soybean and wheat. You can grind the two together. This will ensure a better grinding of the soybean.

Tips for Choosing a Grinder

When you decide to purchase a grinder, these are some qualities to look for:

1. A good warranty—some have two years.
2. Accessibility—choose one that can be used in or near the kitchen.
3. Strong, heavy-duty motor—you will need at least a one-half horsepower motor. A three-fourth horsepower motor is recommended for day-in, day-out grinding.
4. A handle for grinding in case of a power failure. Some have an adapter sprocket that can be operated by a bicycle, water wheel or gasoline engine.
5. Easy adjustments for cracked wheat to fine for baking.
6. Removable drawer or container for convenience in working around the kitchen and for cleaning.

Choosing The Wheat You Buy

Of the leading fourteen species of wheat grown in the world, seven are grown in the United States. The most important of these are common, club and durum wheat. The common wheat is the type needed for our purpose here for it is the bread wheat. From this information, you need only follow a few simple rules to successfully purchase and store wheat.

First, purchase hard wheat with a protein content of from 12 to 15 percent. Most of the protein in wheat is gluten. Gluten is necessary for good bread. The higher the protein content the greater the gluten quantity. Scientists tell us that of 100 grams of typical wheat from the field, 12.5 will be water, 12.3 protein, 71.7 carbohydrates, 1.8 fat, and 1.7 ashes. These relationships vary slightly from one crop and variety to the next.

Second, if the wheat is to be stored for any length of time, the moisture content should be less than 10 percent. This is very important because weevils and insects cannot reproduce nor live in wheat with that low moisture. It will also grind better in your mill.

Third, the wheat should be clean and free of dirt, sticks, grass, rocks and insects.

Storing and Preserving Wheat

Store wheat in clean, dry containers with tight lids. If the lids are not tight, run tape around the edge to get a good seal. Metal containers are often used. Containers holding 50 to 100 pounds of wheat are a good size. Containers with flat lids enable them to be stacked, thus requiring less floor space for storage.

To destroy and protect against weevils and insects, pour an inch or two of wheat in the bottom of the storage container. Place 4 to 8 ounces of dry ice, depending on size of container, on top of the wheat. Continue to fill the container to the top with additional wheat. Allow four to five hours for the ice to evaporate, then seal the lid. Place the containers on wooden shelves or wooden slats to prevent moisture from collecting on the bottom. The healthiest and best storage is rotation by cooking with wheat.

The amount to store depends largely on age, sex and activity. A suggestion is 70 pounds for a small child to 365 pounds per adult for one year. Other basic foods you may want to add:

Powdered milk - 100 lbs. per adult person

Sugar or honey - 100 lbs. per adult person

Salt - 5 lbs. per adult person

Seasonings

Chapter 4

COOKING HELPS

Before we discuss techniques and recipes for cooking with wheat and gluten, let's list a few items you will need for occasional reference.

Abbreviations

The following standard abbreviations are used throughout this book:

tsp.	teaspoon
tbsp.	tablespoon
c.	cup
pt.	pint
qt.	quart
pkg.	package
oz.	ounce
lb.	pound
sm.	small
med.	medium
lg.	large
ingred.	ingredients

All measurements are level. Use standard cup and spoons for measuring.

Metric Conversions of Kitchen Measurements

With the metric system being used with increasing frequency in the United States, this conversion table will be a useful item to you.

Our Present Measurements	*Metric Measurements*
1 teaspoon	5 milliliters (ml)
1 tablespoon	15 milliliters (ml)
1 cup, liquid	0.2366 liter
1 cup, dry	0.275 liter (rounded)
1 pint, liquid	0.4732 liter
1 pint, dry	0.55 liter (rounded)
1 quart, liquid	0.9463 liter

1 quart, dry — 1.10 liter (rounded)
1 pound — 453.59 grams
1 inch — 2.24 centimeters (cm)

Metric Temperature Conversions

Centigrade (C)   Fahrenheit (F)

$$C = 5/9 \, (F-32°)$$

$$F = 9/5 \, (C + 32°)$$

Boiling Point of Water   212°F   100°C
Freezing Point of Water   32°F   0°C

Oven Cooking Temperatures

The following classifications of oven temperatures will make many recipes more easily understood.

Electric Stove	Gas Stove	Classified
225°-250°	0-1/2	cool
250°-275°	1/2-1	very slow
275°-300°	1-2	slow
300°-350°	2-3	very
350°-375°	4	moderate moderate
375°-400°	5	moderately hot
400°-450°	6-7	hot
450°-500°	8-9	very hot

To convert °F to °C, subtract 32° and multiply by 5/9.

Adjusting for Altitude Differences

Bread baking techniques in this book are described for use in the mid-western states. If you are not pleased with your bread and for much different altitude, follow these simple rules:

1. For higher altitude use less yeast, a little less rising time and increase oven temperature a few degrees.

2. For lower altitude add more yeast and allow more rising time.

Ingredient Substitutes

You may wish to substitute one ingredient for another in the recipes with which you are working. The following may be considered "safe" substitutes:

2 c. baked ground gluten = 1 lb. ground round as used in recipes in this book.

1 c. fresh fine ground whole wheat flour = 1 c. sifted all purpose flour for most baking.

3/4 c. fresh fine ground whole wheat flour plus 1/4 c. cornstarch = 1 c. cake flour.

2 tbsp. whole wheat flour = 1 tbsp. cornstarch for thickening.

4 tbsp. dry whole milk plus 1 c. water = 1 c. whole milk.

1 c. whole or skim milk plus 1 tbsp. lemon juice or vinegar = 1 c. buttermilk or sour milk.

2 tbsp. powdered whole egg plus $2\frac{1}{2}$ tbsp. water = 1 whole egg used for baking.

1 c. honey = 3/4 c. granulated sugar plus 1/4 c. liquid. (In cakes only substitute honey for 1/2 of sugar.)

1 sq. unsweetened chocolate = 3-4 tbsp. cocoa plus 1/2 tbsp. shortening.

1 tsp. baking powder = 1/4 tsp. soda and 1/2 tsp. cream of tartar.

*Sprouted Wheat
and
Sprouted Wheat
Cereal*

*Cooked Cracked
Wheat Cereal*

*Whole Grain Wheat
and
Wheat Stalk*

Section One
USING WHEAT

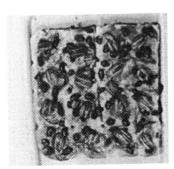

Rolls From Versatile
Whole Wheat Bread Dough
With Syrupy Frosting

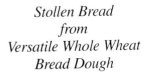

Rolls From Versatile
Whole Wheat Dough

Stollen Bread
from
Versatile Whole Wheat
Bread Dough

Whole Wheat, Rye
and
French Bread

Chapter 5

COOKING WHOLE, CRACKED AND BULGAR WHEAT

Whole Wheat

Three Methods For Cooking Whole Wheat

Here are three different ways to cook whole wheat, based on the following quantities:

    1 c. wheat kernels   1 tsp. salt
    2 c. water

Method 1. Wash wheat, soak in salted water overnight. Boil in same water 5 minutes, simmer 1 to $1^{1}/_{2}$ hours or until tender. It will about triple in volume.

Method 2. Wash wheat, add to boiling salty water. Reduce heat, simmer 5 to 6 hours

Method 3. Wash wheat, add to boiling salty water. Boil 2 minutes, pour in preheated thermos. Seal immediately, let set overnight. Make sure thermos is large enough for wheat to expand.

Two Tasty Ideas For Using Cooked Whole Wheat

Cooked wheat may be drained well, then sprinkled with Worcestershire sauce, Parmesan cheese, garlic, onion or celery salt. Spread thin on baking sheet, roast slightly in 250 degree oven 10 to 20 minutes. Serve as treats.

Cooked wheat may be drained well and dropped in hot fat to brown. Remove to paper towel, seasoned as above. Eat by themselves or add to dried seeds, dried fruit or popcorn.

Wheat Kernel Casserole

    1 c. wheat, precooked, drained   2 c. carrots, diced
    1 lb. ground meat   1 can cream of
                                              mushroom soup

1 lg. onion, chopped
2-3 tbsp. oil
1 c. celery, chopped
2 c. water
2 c. potato chips, crushed
1 lg. pimento, diced

Brown wheat, meat, onions in oil and drain. Add remaining ingredients except potato chips. Butter bottom of baking dish, sprinkle with cup of chips. Pour in mixture, sprinkle other cup of chips on top. Bake 350 degrees for 45 minutes to 1 hour or until vegetables are done.

Add precooked wheat to casseroles, chili, spaghetti sauce, sloppy joes, soup, or may be eaten as breakfast cereal.

Hot Wheat Drink

In heavy, small saucepan, place 2 to 3 cups wheat. Parch to dark brown on medium heat. Stir 3 to 4 times. Crack in blender or mill. Store in closed container on shelf. Use 1 rounded teaspoon to each cup of boiling water. Set about 10 minutes. Strain. Serve plain or with cream and sugar.

Cracked Wheat

You can crack wheat in your wheat mill by adjusting the stones to open, then sifting the wheat. Cracked wheat may also be Purchased in food stores.

Cooked Cracked Wheat

1 c. cracked wheat
3 c. water
1 tsp. salt
1 tbsp. margarine

Combine, cover and bring to boil. Reduce heat and simmer 20 minutes. Serve as hot cereal with milk and sugar or honey. Raisins (1/4 cup) may be added while cooking. About six 1/2 cup servings.

Cracked Wheat Casserole

1 can tuna
1 c. celery, diced
1 med. onion. chopped
1 med. green pepper, chopped
1/4 tsp. pepper
1/3 c. pimento, diced
1/2 tsp. rosemary
1 small can mushroom stems & pieces

3 c. cracked wheat, precooked	1 can cream of mushroom soup
1 tsp. salt	

Drain oil from tuna in fry pan, add celery, onions and green pepper. Cook to tender but not brown. Add tuna and remaining ingredients. Pour in buttered baking dish, bake 350 degrees for 45 minutes.

Add 1/4 cup or more precooked cracked wheat to toss salads, or breads. Add to casseroles, soup and stews.

Bulgar Wheat

Preparing Bulgar Wheat

Wash wheat, cover with water, steam 45 minutes and drain. Spread thin on baking sheet. Place in 250° oven until very dry. Remove from oven and cool. Lightly dampen hands, remove chaff by rubbing kernels between hands. Crack wheat in blender or crack in mill as cracked wheat. Store on shelf in container with loose fitting lid.

Bulgar Cereal

1 c. bulgar	1 tsp. salt
1 tbsp. margarine	$3\frac{1}{2}$ c. water

Add bulgar and margarine to boiling salty water. Simmer 20 minutes. About five 1/2 c. servings.

Bulgar wheat adds flavor when used as part meat in meat loaves, meat balls, chili, stew, soup, curry chicken. and casseroles.

Chapter 6

SPROUTING WHEAT

Sprouting Directions

Wash 1/2 cup untreated wheat. Cover well with tap water and soak overnight. Place wheat in a colander or strainer, mesh tray, any type large bowl, crockery, quart jar, large sponge, etc. Each will need to be covered: a towel, plate or lid will do.

Three or four times a day, hold the container under tap water and rinse. Gently turn wheat with hands. An easy way to remember is rinse them at breakfast, lunch, dinner and before going to bed. Drain well, keep damp (but not wet or they will sour). Keep container covered and near sink for convenience in rinsing. If using a quart jar or see-through container, place it in a dark area or cover it with a towel.

Begin using sprouts when they are the length of a wheat kernel (after about 3 days). To retard sprouting, place the seeds in a closed container in the refrigerator for eight to ten days.

Other seeds and grains may also be sprouted in the same manner. Sprouted wheat enhances the flavor of most foods. Use sprouts by adding 1/2 cup or more to bread, muffin or waffle dough, casseroles, salads, gravy, soup, stews, omelets, sandwiches, etc. Studies made on wheat sprouting show that wheat sprouts are low in calories, the starch is converted to simple sugar for quick energy pick-up, and protein and vitamins are increased.

Sprouted Wheat Cereal or Treat

1 c. sprouts, 3 days old
1/3 c. water
2 tbsp. oil
1/2 tsp. baking powder
1 tbsp. sugar

Blend together in blender. Spread thin on greased baking sheet. Bake 350 degrees 10 to 20 minutes or until brown. If necessary, remove edges when brown to keep from burning. Serve as prepared cereal or treats. Has taste similar to Grapenut flakes.

Wheat Grass

Spread 2 inches of good soil in a 12 x 12 container. Dampen the soil well and sprinkle with 1/3-1/2 cup wheat. Cover with thin layer of soil and pat gently with the hand. Cover with wet cloth. Keep cloth moistened 3 or 4 days, then remove and dampen soil. The container should be kept in a warm room. It is ready to cut and use in 7 to 10 days.

Add to drinks made in blender or cut and add to soups and stews.

Chapter 7

BREAD MAKING WITH WHOLE WHEAT FLOUR

Discover how easy, fun and tasty it is to make yeast and plain breads. All you do is follow a few, simple rules.

Basic Bread Ingredients

These are busy days so have one basic bread recipe that can be used for making different kinds of breads and rolls. Learn about the necessary ingredients and what they do. This will enable you to adjust recipes for your specific taste. Take a look at each ingredient involved in breadmaking.

1. *Flour* is the basic structure of bread. Use 100% stone ground (freshly ground) flour or sift other flour twice for accurate measure. If you grind your own flour, use hard wheat which is 12-15 percent protein. This contains more gluten, the special substance needed for making good bread. As the dough is kneaded or pounded gluten forms an elastic-like shell for the gas bubbles which yeast produce.

   Wheat flour produces light bread. Other flours (rye, buckwheat or soy) may be used for variety on a safe ratio of one Part to 3 parts whole wheat flour.

2. *Liquids* may be milk, water, potato water, water from washing gluten, or equal parts of water and vegetable juice or tomato juice. Water gives bread a crusty crust and more wheat taste. Milk gives bread a softer crust and texture. Liquids need to be lukewarm. If milk is used it must be scalded.

3. *Yeast* manufactures gas that causes the bread to rise. There are two kinds:

   Dry yeast in granular form. It is packaged in small foil packages or in metal containers in larger quantities. Storage time is up to one year in refrigerator.

Compressed yeast is in square cake form. It must have immediate refrigeration. Its life is about 2 weeks. It will retain its potency about 2 months when frozen.

Usually, in making bread, yeast is added to a small amount of warm water to dissolve, which takes about 10 minutes. One tablespoon of sugar added to it will hasten its dissolving. Water which is too hot will kill yeast. A yeast taste in bread is from water which is too hot, not from the amount of yeast used.

4. *Sugar,* granulated or brown, honey or molasses, is food for the yeast. It also adds color and flavor, and aids in browning as the bread bakes.

5. *Salt* enhances flavor and works on the yeast to retard gas formation.

6. *Fats* may be oil, shortening, butter, margarine or lard. Fats give bread a soft, tender texture. For some types of bread such as French bread and hard rolls, less fats are used.

7. *Eggs* are sometimes used to increase food value, flavor and produce finer texture.

8. *Other ingredients* such as raisins, dates, prunes, figs, nuts, spices, seeds, dry onion soup, etc.. may be added.

Kneading or Pounding Dough

For speed and ease, the ideal way to knead bread is with an electric bread kneader. Today there are mixers on the market that have kneading arms for kneading breads and rolls as well as beater whips for ordinary mixing.

Electric Mixer Method

Directions for kneading dough are included with the instructions for most electric mixers. The following are general guidelines you can use if the instructions are unavailable.

1. Sprinkle yeast over 1/2 cup warm water in small bowl. Do not stir. Set aside to work. One tbsp. of sugar sprinkled on the yeast will help it work faster.

2. With the bread kneader mix liquid and 1/2 of the flour in the mixer bowl.

3. Add sugar, salt, oil, and mix. Add 1 cup flour to mixture.

4. Add prepared yeast to mixture and mix well.

5. Add the rest of the flour and knead for the time specified in the recipe (Typical times: eight to ten minutes for bread, about five minutes for sweet rolls and fancy breads). Stop the mixer and scrape the bowl with a spatula if necessary.

6. *Caution:* You can overload your machine's capacity when making bread. Don't use too high a speed for kneading. Don't use too large a recipe. Experiment and learn the capacity of your machine.

7. A bread mixer can be used for any recipe requiring kneading. Experiment with your favorites and enjoy your investment. (It is an excellent tool for beating fondant in dipping chocolates. for mixing fruit cake granola, etc.)

If you do not have access to an electric kneader, use the following simple pounding method (rather than hand kneading, which requires one to three rising periods). Mix according to instructions given in recipe.

Pounding Method

1. Use an 18 x 22 inch canvas bag or one made from material that is plastic lined (1/2 yard of 45 inch material is sufficient).

2. After dough has been mixed, place it into the bag. Twist the open end and hold it with one hand.

3. Place it on the bread board and pound it five minutes with the broad side of a carpenter's hammer or similar hammer, a large wooden rolling pin a club or similar object.

4. With greased hands remove the dough to a flat, oiled surface.

5. To get even-sized loaves, pinch or cut the dough in half, then in fourths, etc., depending on the amount of dough and the size of the loaf pan.

Forming Loaves in Bread Pans

To form the loaves, flatten the dough with greased hands into a rectangular shape. Fold the two short sides to the middle and press out air and gas bubbles. Lift the dough from opposite sides with both hands and slap the center bottom on a kneading surface. This lengthens the dough. Fold each end you are holding to the center again, pressing out air and gas bubbles as you do it. Roll the dough tightly from one side to the other, sealing seams and pinching the ends to seal the loaf. Place the dough seam-side down in a greased loaf pan and press it to fit the pan.

Grease the pans to be used for baking with shortening. Any size or shape pan may be used. This includes regular loaf pans in metal, Pyrex, Corning ware or foil. Juice cans, flower pots—in fact, any type utensil that can be put in your oven—will do to bake bread in.

Rising time varies according to the temperature of the dough and the room and also according to the amount of yeast used. Bread is usually ready to bake when fingertips pressed in the dough leave only a slight indentation and the loaf is well-rounded. Oven temperatures will vary; follow instructions in recipes.

The general appearance of a good loaf of bread is symmetrical in shape. A slice cut from the center of the loaf should differ very little in size from those cut near the ends. A well-shaped loaf will have a well-rounded, smooth dome with no bulges. The bottom and sides of the loaf should be a rich golden brown with the top slightly darker. The texture should have a velvety feel with no hard spots, and with small uniform cells.

Causes of Poor Quality Bread

Several problems can occur to reduce the quality of home cooked bread. These problems can be easily corrected if you understand their causes. If a loaf cracks on the side, the problem is usually one of three things: (1) more raising time is needed, (2) the oven heat is uneven, or (3) the pans are too close together in the oven.

Lumps and bulges on top of the loaf may occur if the crust separates from the rest of the loaf. This may be the result of bread
rising too long and also getting too light. Getting too light will cause a thick crust also. It may also be from poor methods of forming dough

into loaves or from leaving a gas bubble just under the top of the loaf when the loaf is formed.

Crumbly bread is caused by too much flour being used. Bread will also crumble if the rising time of the loaf is too long, causing it to get too light.

Course, uneven texture in bread comes from too little kneading or too long a rising period.

Tips For Using Old Bread

Try these techniques for using day-old bread:

Bread Crumbs: make crumbs from day-old bread by blending the bread in a blender, or by putting it through a food chopper.

Croutons: remove crusts and slightly butter slices of day-old bread on both sides. Cut the bread in tiny squares, and toast it in the oven about 20 minutes.

Melba Toast: slice day-old bread in 1/8 to 1/4 inch slices. Sprinkle it with garlic or onion salt, with cinnamon, or keep it plain. Bake it at 250° for 20 to 25 minutes or until it is crisp and light brown.

Crisp Loaf: slice a loaf of bread. Keep the loaf intact but spread the slices apart at the top. Pour melted butter over the top. Heat uncovered in a 350° oven for 10 to 15 minutes.

Basic Recipe

Versatile Whole Wheat Yeast Bread

2 tbsp. dry yeast
1/2 c. warm water
5 c. warm water
12-13 c. whole wheat flour, freshly ground or sifted twice
2/3 c. salad or cooking oil
2/3 c. honey or molasses (or may use 1/2 honey and 1/2 molasses
2 tbsp. salt
4 eggs, optional (the eggs are necessary in all following recipes except Rye Bread)

If bread kneader is used, follow instructions for mixing and kneading. If mixing by hand, sprinkle yeast over 1/2 c. warm water, set aside. When dissolved, the mixture should rise to top of cup. In a large mixing bowl combine water and one half flour. If eggs are used, add to batter and mix well. Add oil, honey, salt, and mix in. Stir in yeast mixture, and add enough of remaining flour to form a ball. Place dough in bag and pound 5 minutes with the broad side of a carpenters hammer, turning the bag several times while pounding. Form each loaf to fit pan. Let rise to well-rounded loaves, about 1/3.

Bake 450 degrees 10 minutes, then 350 degrees 25 minutes. If softer crust is desired, bake 350 degrees 35 to 40 minutes. Remove bread from pans and grease top with butter, cover until cool.

Rye Bread

>1/4 recipe of dough 1/2 tsp. Anise seed
>1 tbsp. Caraway seed

Knead together just until mixed, in bread kneader or by hand. Form in two long rolls. Place in greased rye bread pans or on cookie sheet. Let rise and bake as directed. Makes 2 loaves.

Sliced, Buttered Bread Ring

Roll out 1/4 dough to 1/2 inch thickness. Cut with biscuit cutter. Dip each slice on both sides in melted butter. Arrange slices edge ways around tube pan. Let rise and bake as bread. Makes 1 ring.

Crispy Buttered Rolls

Prepare buttered slices as for Sliced Buttered Bread Ring. On greased cookie sheet place slices overlapping each slice halfway. Bake 450° oven about 20 minutes. Makes 25 to 30 rolls.

Stollen Bread

>1/4 dough recipe 3/4 c. raisins
>1 tsp. vanilla 3/4 c. candied mixed fruit or cherries
>2 tbsp. orange rind, 1/2 c. almonds, chopped (or pecans
> grated   or walnuts)

Bread Making with Whole Wheat Flour

Knead together all ingredients until well mixed. Divide dough in half. Place each half on oiled surface and with buttered fingers pat each to size of a dinner plate. Fold each over halfway allowing the top fold to extend beyond the lower fold. Press edges firmly so it will not open. Let rise about double. Bake 425 degrees 20 to 25 minutes. Frost with powdered sugar frosting. Makes 2 large loaves.

Stick Bread

Divide 1/4 of the dough into 20 balls. Roll each ball between your palms to resemble a stick 1/2 inch thick and the length of the loaf pan. Now roll each stick in melted butter and place equal amounts in two small loaf pans. Let rise. Bake 450 degrees 10 minutes then 350 degrees 25 minutes. Makes 2 small loaves.

Dill Bread

1/4 bread dough recipe	2 tsp. dill seed
1 tbsp. lemon juice	1 tbsp. dry onion soup

Knead together just until mixed well. Form into loaves. Bake with rest of bread. Makes 1 large or 2 small loaves.

Lemon Nut Bread

1/4 bread dough recipe	Grated rind of 2 lemons
1 c. nuts, chopped	Glaze

Knead together just until mixed well. Form into loaves. Bake with rest of bread. While still warm, prick bread with fork. Pour glaze over loaves made with 1 cup powdered sugar and juice of 2 lemons. Makes 1 large or 2 small loaves.

Donuts or Scones

1/4 bread dough recipe	Oil for deep frying

Roll out dough to 1/3 inch thickness. Using $2^{3}/_{4}$ inch cutter. Cut out donuts, or with a knife cut small rectangles for scones, let rise to light. Fry in deep, hot oil. Glaze at once using 1 pound box powdered sugar, 3/4 c. boiling water, and 1 tsp. vanilla. Mix together and dip each donut. Makes about 3 dozen.

Crunch Top Rolls

1/4 bread dough recipe
4-5 tbsp. melted butter or margarine
1/8 tsp. garlic or onion salt, optional
1/2 c. cornmeal

Form dough into about 48-1 inch balls. Roll each ball in melted butter, then in cornmeal to which garlic or onion salt has been added, if desired. Place in dripper pan, leaving a little space between each ball for rising. Let rise. Bake 450 degrees 10 minutes, then 350 degrees 10 minutes, or until brown. These can be baked with regular loaves of bread. Makes about 4 dozen.

Breakfast Cake

Roll or pat with hands 1/4 of the dough to 10 x 15 inch rectangle. Brush with 1/2 c. melted butter or margarine. Sprinkle with one cup brown sugar and 2 tablespoons cinnamon. Roll dough as for jelly roll from long side. Place roll, seam side down in greased tube pan. Seal ends together. Let rise in warm place to double, about 35 minutes. Bake 350 degrees about 45 minutes. Turn out on serving plate and cover top and sides with breakfast cake frosting.

Breakfast Cake Frosting

2 tbsp. butter or margarine
1 c. powdered sugar
1/4 c. brown sugar
2 tbsp. light cream
1/2 c. nuts, chopped

In small saucepan melt butter, add brown sugar and cream, add powdered sugar. Beat until creamy. Frost rolls or bread, sprinkle with nuts. Makes about 1 cup.

Sweet Rolls

Roll or pat with hands 1/8 of dough into rectangle about 8 x 12 inches. Brush with 2 tablespoons melted butter, sprinkle with 1/2 c. brown sugar then 1 tablespoon cinnamon. Roll dough from long side, seal edges. Slice into 16 slices and arrange over one of the following frostings in a 9 inch square pan. Let rise in warm place to double, about 35 minutes. Bake 425 degrees 20 minutes. Makes 16 rolls.

Syrupy Frosting

> 2 tbsp. butter or margarine 1/4 c. dark corn syrup
> 1/2 c. brown sugar 1/2 c. nuts, chopped
> 1/4 c. raisins

In a 9 inch square pan, melt butter, add sugar and syrup, and heat just until sugar dissolves. Sprinkle with nuts and raisins. Arrange rolls on top. Makes 1 cup.

Nuts and Cinnamon Frosting

> 2 tbsp. butter or margarine 1/4 c. light cream
> 2 tsp. cinnamon 1/2 c. nuts, chopped
> 3/4 c. powdered sugar

In 9 inch square pan melt butter. Stir in sugar, cream and cinnamon. Sprinkle with nuts. Arrange roll slices. Makes about 3/4 cup.

Nuts and Powdered Sugar Frosting

> 2 tbsp. butter or margarine 3-4 maraschino cherries, chopped
> 3/4 c. powdered sugar 1/2 c. nuts, chopped
> 1/4 c. light cream
> 1/4 c. crushed pineapple, optional

Mix and use same as Nuts and Cinnamon Frosting. Makes about 1 cup.

Cinnamon Bread

Roll or pat with hands 1/4 of dough the length of loaf pan and 10-15 inches wide. Brush with 1/2 cup melted butter. Sprinkle with 1 to $1\frac{1}{2}$ Cups brown sugar and 2 or 3 tablespoons cinnamon. Roll up from short side. Seal edges and ends. Place seam side down in greased loaf pan. Let rise in warm place to double size, about 45 minutes. Bake as you bake rest of bread or rolls. When cool slice, lightly butter, toast under broiler, and serve for breakfast. The slices may also be put through food chopper on medium blade, dried out in slow oven and served as breakfast cereal.

Whole Wheat French Bread

2 tbsp. dry yeast	2 tbsp. shortening
1/2 c. warm water	1/4 c. sugar
1 qt. warm water	$11\frac{1}{2}$ c. freshly ground
4 tsp. salt	or sifted whole wheat flour

If bread kneader is used follow instruction for mixing and kneading. If by hand, sprinkle yeast over one-half cup warm water, set aside to dissolve. In large bowl combine warm water, salt, shortening and sugar. Stir in 2 cups flour, add yeast mixture. Stir in rest of flour to form dough. Place dough in bag and pound 5 minutes with the broad side of a carpenters hammer, turning the bag several times while pounding. Divide dough in 2 equal parts, roll each to 9 x 12 inch rectangle. Starting at one corner roll diagonally to opposite corner, seal edges, place seam side down on well-greased baking sheet sprinkled with cornmeal. Allow space for loaves to rise. Cut 3 diagonal slits into top of each loaf. Let rise to double in bulk, about 1 hour. Brush top and sides with slightly beaten egg white, sprinkle with sesame seeds. Bake 450 degrees 10 minutes. Reduce heat to 350 degrees for 25 minutes. This recipe may also be used for hard rolls. Use 10 minutes less baking time. Makes 2 long loaves or about 20 hard rolls.

Whole Wheat Rolls or Buns

$3\frac{1}{2}$ c. warm water	1 tbsp. salt
3/4 c. oil	3 eggs, beaten
1/2 c. sugar	$10\frac{1}{2}$ c. whole wheat flour
6 tbsp. dry yeast	freshly ground or sifted twice

In large mixer bowl mix water, oil, sugar and yeast at medium speed. Let rest 15 minutes. Add salt, eggs and half of flour. Beat well. If using rotary mixer it may be necessary to beat in rest of flour with large spoon. Shape immediately in desired rolls or buns. Let rise 15 minutes. Bake 425 degrees 12-15 minutes. This dough makes good cinnamon rolls, dinner rolls, hot dog buns, hamburger buns, donuts and scones.

Hot Dog Buns

Form dough with hands, give each bun a twist as they are placed on the baking sheet.

Fancy Rolls

Make long rolls of dough 1/2 inch thick. Cut roll in 6 inch lengths, tie in knots, or let children form their initials. Dip in milk, sprinkle with cinnamon or orange sugar or fine chopped nuts. Make nut twist by cutting roll in 8 inch length, dipping in milk and fine chopped nuts, and forming figure eight as you place on greased cookie sheet. Braids and sticks may be formed. To form cloverleaf rolls cut roll into pieces to make 1 inch thick ball. Place 3 balls in each cup of greased muffin pan, brush with butter.

For Crescent Rolls, roll ball of dough in circle 1/4 inch thick. Cut pie-shape wedges, brush with melted butter, roll up from wide end. To form Parkerhouse Rolls, roll dough 1/4 inch thick, cut with 3 inch biscuit cutter, make dent just off center, brush with butter, fold over so top side is the larger portion, brush top with butter.

Basic Recipe

Basic Wheat Mix [BWM] for Quick Bread

12 c. whole wheat flour sifted
1/2 c. baking powder
$2\frac{1}{2}$ c. non-fat dry milk
5 tbsp. sugar

3 c. shortening, use part margarine
$1\frac{1}{2}$ tbsp. salt

In large bowl sift dry ingredients together, cut in shortening. Store in covered container in refrigerator. Keeps well up to 3 months. (A wide-mouth gallon glass jar is good for storing.) Use in following recipes.

Biscuits

2 c. BWM   2/3 c. water

Mix together to soft dough. Sprinkle smooth flat surface with BWM, gently knead dough 8-10 times. Flatten dough to 3/4 inch thickness. With knife cut in 12 squares. Place on ungreased baking sheet, brush with melted butter. Bake 425 degrees about 15 minutes or until nicely brown.

465Biscuit Variations

Use same ingredients, mix as for biscuits, adding any one of the following:

2 tbsp. dry onion soup
or
1/4 tsp. nutmeg, 1/4 tsp. sage, and 1 tsp. caraway seed
or
1/2-3/4 c. grated sharp cheese
or

Dip biscuit in melted butter. Roll in mixture of 1/2 cup sugar, $1\frac{1}{2}$ tsp. cinnamon. Sprinkle with nuts if desired

Pancakes

1 egg   2 c. BWM
$1\frac{2}{3}$ c. water

Blend together in blender. Pour on heated griddle. For thinner batter add more water. Makes about 18 medium pancakes.

Waffles

1 egg $1\frac{2}{3}$ c. water
2 tbsp. cooking oil   2 c. BWM

Blend together in blender. Pour on heated irons. Makes 3 large waffles.

Dumplings

1 egg, beaten add
Water to make 3/4 c.
2 c. BWM

Mix together to form stiff dough. Roll to 1/8-1/4 inch thickness. Cut in 1-inch wide strips. Drop in 4-6 cups boiling broth. Boil 10-12 minutes. Stir gently, keep dumpling punched down in broth for even cooking.

This dough works nicely for meat and chicken pies and pizza dough. It may also be used for a quick fruit cobbler on top of stove. Use 1 quart fruit—peaches, berries, apricots, plus 1-2 cups water. Roll, cut dough and cook same as for dumplings.

Section Two
USING GLUTEN

Sauce For Spaghetti

2 Tasty Meat Loaves In One Mix

Sausage Skillet Dish

Quick Make Ahead Tuna Dish

Special Franks Versatile Meat Balls

Hawaii Chicken With Meat Balls

Baked and Ground Gluten

Chapter 8

GLUTEN: EASY TO MAKE, EASY TO USE

Creative Wheat Cookery can be a treasured addition to your kitchen and food storage library by helping you extend wheat beyond breads and cereals through the use of Gluten.

You'll find that this book explains everything you need to know about gluten: what it is, its versatility and economy of its use, and its nutritional value. Easy instructions for making and using many tasty items, plus simple ways to adapt to your favorite recipes, and also many brand new kitchen-tested recipes that can be used everyday, are included.

Advantages of Cooking with Gluten

Gluten is a protein source. Gluten made from whole wheat flour is an excellent source of protein, especially when supplemented with milk, cheese, eggs, meat, fish, poultry, soybeans and other beans.

Gluten is versatile. In raw form it is tough, has an elastic texture, is almost tasteless, and is insoluble in liquid. It can be used effectively as a substitute for meat. Discover quick, easy methods in preparing mock steaks, wheat balls and patties, hamburgers, franks, sausages, tasty casseroles, delicious salads, easy desserts and pet food.

Be thrifty. Add nutrients by becoming skilled in using the starch, water and bran left after extracting gluten. You can be successful the first time by following simple, easy instructions.

Gluten is economical in that it costs about 1/4 the price of ground round. Use gluten with seasoning for tasty meals, or use part gluten and part ground round, chicken, ham or tuna.

Gluten stores easily at any stage. Raw, baked or ground gluten will store indefinitely in a freezer. It may be thawed and refrozen many times. After being baked and ground it may be completely dried in the oven and stored in a closed canister indefinitely. Raw gluten will store in your refrigerator about a week, while baked or ground gluten will keep about 10 days. Because it stores well it is very convenient to use.

Gluten yield from high protein hard wheat flour is higher than from soft low protein flour. High protein whole wheat flour also yields more nutrition and better quality gluten than white flour. Gluten used in the recipes of this book was made from high protein wheat (15 percent).

Calculating Gluten Quantities

The following will help you in determining yields when gluten is being made from wheat and when gluten is being combined with other foods:

> 16 cups whole wheat flour plus 8 cups water yield about 5 cups raw gluten, 4 to 5 cups bran and 2 to 3 cups starch water.
>
> 5 cups raw gluten, baked and ground, yield 10 to 11 cups ground gluten.
>
> 2 cups of ground gluten is equal in bulk to 1 pound ground round as used in recipes in this publication.
>
> 1 cup raw gluten contains 72 grams of protein.

A simple rule to remember when using gluten is to let 10 to 25 percent of the total bulk of the dish be other protein foods such as milk, cheese, eggs, meat, chicken, tuna or legumes. Most of the recipes in this book using gluten as a main dish have followed this simple rule.

It is easy to supplement gluten with other foods for a complete protein dish.

> Some suggested supplements are as follows:
>
> 4 tbsp. textured vegetable protein (T.V.P.), dry, contain 10 grams of protein
>
> 1 egg contains 6 grams of protein
>
> 1 oz. hamburger contains 7 grams of protein
>
> $1\frac{1}{2}$ c. cooked beans contain 7 grams of protein
>
> 1 c. liquid milk contains 9 grams of protein
>
> 1 oz. Process American Cheese contains 7 grams of Protein

Extracting Gluten in Three Easy Steps

Step I: Knead the Wheat-Water Mixture:

Given below are three alternate methods to make gluten in 4 minutes or less, which are easy and real time savers. Method 1 uses an electric bread kneader. Method 2 uses an electric rotary mixer. Method 3 is done by hand.

Method 1: Using an electric bread kneader: Use 1 part tap water to 2 parts freshly-ground or sifted whole wheat flour. For a beginner use 4 cups water to 8 cups flour. Combine water and flour in kneader bowl and knead 4 minutes. Add 1 qt. of water and knead 1 minute. This quickens the separating of the gluten from the starch and bran. This method is the ideal way—it is faster and easier. Proceed to step II.

Method 2: Using a standard electric rotary mixer: Combine 3 cups tap water and 4 cups freshly-ground or sifted whole wheat flour in mixer bowl and mix 4 minutes at medium to high speed. Scrape side of bowl often. Caution: Motor may over heat. If so, rest it by beating one minute, resting 2 minutes, etc. Proceed to step II.

Method 3: By hand. Use same proportion of water and flour as in method 1. Mix water and flour in a large mixing bowl to form the dough. Place the dough in a heavy wet plastic bag or in a wet canvas bag. Twist the open end and tie it with a metal tie or rubber band. Hold the twisted end with one hand and, with the broad side of a carpenter's hammer or a hammer of equal width and weight, pound the dough 4 minutes turning the bag several times during the pounding. Remove dough from the bag to the large mixing bowl. (Material For Bag: 1/2 yd. 45" width. Place selvage together. Sew sides rounding at bottom so as not to have corners. Use seam binding to reinforce seams.) Proceed to step II.

Step II: Separate the Gluten from the Starch and Bran:

Set the bowl with the dough in the sink and fill it with tap water. With both hands squeeze the dough through your fingers until it just begins to feel like it is coming apart in shreds. The water will begin to thicken and look milky. Under slow-running tap water place a colander in a deep pan. Take handfuls of dough at a time and, while holding it under the tap water, squeeze and turn it in your hands to wash out the starch and part of the bran. Continue until it feels as elastic as chewed bubble gum and a bit grainy with bran. Place the washed, raw gluten in a bowl while you continue to wash the rest of the dough. If you wish, use 2-3 quarts of water in a large bowl instead of running tap water.

Step III: Save the Starch and Bran:

Save the liquid in the deep pan and mixing bowl. In about 30 minutes the bran and starch will settle to the bottom. Pour off the clear water and store it in the refrigerator to be used in soups, stews, gravy, chili, spaghetti sauce, casseroles, bread, and vegetables. This is important because the water contains vitamins and minerals. The starch and bran may be poured in a container, covered, and placed in the refrigerator. Or the starch water can be dipped off and each stored separately for use when convenient. They may be stored a week to 10 days in refrigerator. In the Starch and Bran section are many suggested uses and tasty recipes.

Baking and Grinding Gluten for Quick Use

Ground Gluten: Spread the gluten by patting it on a board to a 1/2 inch thickness. Cut it in long 1 to $1^1/_2$ inch wide strips. Place it on a baking sheet greased well with shortening or on a lightly-greased Teflon baking sheet. Bake 350 degrees 25 to 30 minutes.

To test, press a finger into the baked gluten. When done the gluten will spring back.

The baked gluten may be put through a food chopper using a medium blade. Long strips of baked gluten lets you feed the hopper more easily. The baked gluten may also be shredded on a shredder using larger holes. The ground gluten may be used as ground round. [See *Use Gluten in Place of Meat, Use Gluten with Meat, and Tasty Toppings* sections.]

Mock Steaks: With shortening generously grease foil baby loaf pans; or empty clean soup, tomato sauce or tuna cans, and fill them 1/2 full with raw gluten. Bake 325 degrees about 1 hour. Run a knife around the edge of the container and carefully remove the baked gluten. [See mock steak recipes.]

Lunchmeat: Bake gluten the same as for "Mock Steaks." [See lunch meat recipes.]

Chapter 9

GLUTEN IN PLACE OF MEAT

Ground Gluten as a Substitute for Hamburger

Ground gluten can be used in any recipe calling for ground meat by adding some beef flavor, such as Schilling Beef Flavor Base, Beef Consomme, Beef Bouillon, Bernards Beef Flavor Soup, Spice Island Beef Flavor Base, or other beef flavors, chicken flavor, soups and cheese. Some can be purchased from restaurant supply houses. Use 2 cups ground gluten for each 1 pound ground round needed in your favorite recipes.

Gluten can be very tasty when used in seasoned dishes such as spaghetti sauce, franks, sloppy joes, curry dishes, enchiladas, tamales, meatless balls and patties, lasagna, etc. Use the water from extracting gluten in these recipes where water is needed. If extra coloring is desired, use Kitchen Bouquet or red food coloring.

Best Ever Lasagna

1/2 lb. lasagna noodles
   or broad noodles
3 qts. boiling water
2 tbsp. salt
1 tbsp. butter

1 lb. small curd cottage cheese
1/2 lb. mozzarella cheese sliced
1 tomato sauce recipe
1/2 c. grated Romano cheese

Cook lasagna in boiling salted water until just tender. Drain well and mix with butter. Arrange in a shallow baking dish with layers of each in this order: lasagna noodles, cottage cheese, sliced mozzarella tomato sauce and last, Romano cheese. Bake 325 degrees 40 to 45 minutes

Tomato Sauce:

1 6 oz. can tomato paste
1 lb. canned tomatoes
Pepper to taste
1/4 tsp. garlic pwd.
1/2 tsp. dried parsley flakes

Pinch of basil
2 c. water
1 tbsp. Schilling Beef Flavor
2 c. ground gluten

Mix together all ingredients except gluten. Simmer 1 hour, then add gluten and simmer 20 minutes more. Before you assemble make the sauce. Serves 10-12.

O-So-Yummy Puffs

1 c. ground gluten	1/2 tsp. paprika
1 tbsp. Schilling Beef Flavor Base	1 tbsp. parsley, chopped
3 egg yolks, slightly beaten	2 tbsp. onion, finely chopped
1/4 tsp. baking powder	3 egg whites, stiffly beaten
1/2 tsp. salt	1 c. cooking oil
1/4 tsp. pepper	

Mix first 9 ingredients. Fold in egg whites. Heat oil in 8 or 9 inch fry pan. Drop mixture by teaspoonfuls in hot oil. Brown on both sides. Drain on paper towel. Serve immediately as is or in mushroom sauce from sauce section or with catsup.

Favorite Winter Chili Pot

2 c. ground gluten	4 c. canned tomatoes
3 tbsp. oil or butter	1 3 oz. can tomato paste
2 c. sliced onions	2 tbsp. Schilling Beef Flavor Base
1 tsp. chopped garlic	3 c. cooked red kidney beans
3 tbsp. chili pwd. (less if desired)	(soak overnight before cooking)
1 bay leaf	Dash of cumin

Brown gluten, onions and garlic in oil. Add remaining ingredients and simmer 45 minutes—makes 2 1/2 quarts. Serves 8-10.

Versatile Meatless Balls or Patties

2 c. ground gluten	1 1/2 tsp. salt
2 tsp. Schilling Beef Flavor Base	1/2 tsp. garlic salt
2 tbsp. cooking oil	1/2 tsp. pepper
4 eggs, beaten	1/2 tsp. poultry seasoning
2 c. bread crumbs, finely ground	1 med. onion, finely chopped
	1 c. Chicken with Rice Soup

Gluten in Place of Meat

Mix all ingredients together. Form into about 60 1 inch balls or 20 patties. Brown on top of stove in 1/2 cup cooking oil or bake 350 degrees 20 minutes for balls and 40-45 minutes for patties. Use as you would any meat ball or patty. Suggestion: Serve in one of the sauces from the sauce section or your favorite sauce. Patties can be grilled outdoors, or can be used in a foil dinner for camping- Each can be made ahead of time, baked and then frozen. Take out only amount needed, keep rest frozen. Use as a base for franks, sausage, or poultry dressing.

Wheat Franks

1/2 recipe meatless ball or patties
1/4 tsp. garlic salt
1 lb. pork sausage extender [from Using Gluten with Meat]

Mix together thoroughly. Form into wiener-size links. Bake 350 degrees 30 minutes on greased sheet. Serve as a hot dog with all the trimmings or as a corn dog. Also make patties to grill or fry for hamburgers. These are especially good when served with mustard

Good Corn Dog Batter:

1 egg, beaten
1/2 c. milk
1/4 c. corn meal
1/4 tsp. salt
1/2 c. whole wheat flour or white flout
1/2 tsp. baking powder
1 tbsp. sugar

Mix together all ingredients. Dip franks in the batter and fry in deep hot oil until brown

Wheat Balls

2 c. ground gluten
2 tbsp. onion, finely chopped
2 tbsp. wheat germ
2 tbsp. Schilling Beef Flavor Base
2 tbsp. olive oil or butter or cooking oil
1 egg, beaten
1/4 tsp. pepper
1 tbsp. parsley. minced

Mix together all ingredients. Form about 25 one inch balls Bake 350 degrees 20 minutes or fry in small amount of oil Serve as is for treats, dinner or simmer 15-20 minutes in sauce from Special Sauces. About 25 Balls.

Tasty Baked Beans

2 c. ground gluten
4 tbsp. olive oil or cooking oil
1 lg. onion chopped
2 tbsp. Schilling Beef Flavor Base
1 tsp. prepared mustard
1/2 c. catsup
1 tbsp. molasses
1 tbsp. brown sugar
1 lg. can pork and beans

In fry pan brown gluten and onion in oil. Add remaining ingredients and pour in baking dish. Suggestion: Sprinkle with bacon bits or slightly-cooked strips of bacon placed on top. Bake 325 degrees for 45 minutes. Serves 8-10.

Beans from Scratch with a Vegetable Relish

1 1/2 c. red kidney beans
4 tbsp. oil or margarine
1 c. ground gluten
2 c. onions, chopped
1 c. mild cheese, grated
2 c. Monterey cheese, grated
1 tsp. salt or more
1/2 tsp. pepper
1 tbsp. Schilling Beef Flavor Base
2 tsp. sweet basil leaves

Wash and soak beans overnight in about 5 cups of water. Next morning brown gluten and onions in oil and add to beans. Cook until tender. Add cheese and seasoning. Simmer about 20 minutes. These are good served hot as is or served cold with the following:

Vegetable Relish

2 c. IMO or sour cream
6 or 7 green chives, chopped
1 large cucumber, sliced
1 c. radishes, sliced
1 tsp. dill seed

Blend ingredients in blender. To serve place beans on serving plates, cover with shredded lettuce, pour relish over all. Suggestion: Use as dip for chips, melba toast or raw vegetables, i.e. carrots, turnips, zucchini, etc. Serves 10-12. (Chop vegetables fine. Stir in sour cream—do not blend for dip.

Quick Spaghetti Casserole

2 c. ground gluten
1 med. onion chopped
1 can cream of mushroom soup

Gluten in Place of Meat

3 tbsp. cooking oil
2 tbsp. Schilling Beef
  Flavor Base
1 can Franco-American Spaghetti
  in tomato sauce with cheese

Brown gluten and onion in oil. Add beef flavor, soup and spaghetti. Bake 350 degrees 30 minutes. Suggestion: Add grated cheese or potato chips on top during last five minutes baking time. Serves 5-6.

Spanish Cornbread

1½ c. cream style corn
1 c. yellow corn meal plain
2 eggs
1 c. buttermilk
1/2 tsp. soda
1 tsp. baking powder
3/4 tsp. salt
1/2 c. oil or bacon drippings
2 c. ground gluten
4 tbsp. oil
2 tsp. Schilling Beef Flavor Base
1 lg. onion, chopped
1/2 lb. cheddar cheese, grated
1 small can
  jalapeño peppers (optional)

Mix together the first 8 ingredients for cornbread and set aside. Brown the gluten and onions in 4 tbsp. oil. Add remaining ingredients. Put one-half cornbread mixture in large greased casserole. Spread the gluten mixture evenly over the top. Pour over the rest of the cornbread mixture. Bake one hour at 350 degrees. Suggestion: Very good with pinto beans. Serves 10-12.

Sausage Casserole

2 c. sausage extender
  [from *using gluten as
  part meat* section]
3 c. cooked rice
1 onion, finely chopped
1 can cream of tomato soup
1/4 c. cheddar cheese, grated

Break sausage in small bits and fry in fry pan, adding 1 tablespoon oil if necessary. Set aside. Place one half of rice in bottom of buttered casserole. Sprinkle with onion then make a layer of 1/2 of sausage. Repeat layers. Pour soup on top. Sprinkle with cheese. Bake 350 degrees 30 minutes. Serves 6-7.

Quick Potato Casserole

4-5 potatoes
2 c. ground gluten
1 can cream of mushroom soup
2 tsp. Schilling Beef Flavor Base

3 tbsp. oil
1 can vegetable soup
1 med. onion chopped
Salt and pepper

Peel and cut potatoes into small pieces and place in bottom of buttered casserole. Brown gluten and onions in oil then add soups, salt, pepper and beef flavor. Pour over potatoes. Bake 350 degrees 1 hour or until potatoes are tender. Suggestion: If in a hurry, boil potatoes to near done first. This cuts baking time in half. Serves 5-6.

Oriental Casserole

2 c. ground gluten
1 med. onion chopped
1 c. celery finely chopped
4 tbsp. butter
1/2 c. uncooked rice
2 tbsp. Schilling Beef Flavor Base
1 can cream of mushroom soup
1 can cream of chicken soup
1 can bean sprouts
2 tbsp. soy sauce
$1\frac{1}{2}$ c. water
Dash pepper
Salt to taste
1 tsp. kitchen bouquet (optional)

Slightly brown gluten, onion and celery in butter. Add remaining ingredients and mix well. Pour in buttered casserole and bake covered 30 minutes in 350 degree oven. Remove cover and bake 30 minutes longer. Serves 8-10.

Sauce for Spaghetti

1/4 c. margarine
1 lg. onion, chopped
3 cloves garlic, chopped
2 c. ground gluten
3 cans tomato sauce
3 cans tomato soup
1 can beef stock
2 tsp. Worcestershire sauce
Shake of Tabasco Sauce
1/4 tsp. pepper
1 tbsp. salt
1 lg. carrot, chopped
1 rib celery, chopped
1 bay leaf
1/2 tsp. oregano
1/2 tsp. rosemary
1/4 tsp. thyme

Saute onion and garlic in margarine. Remove from pan. Add gluten and brown, adding a little oil if necessary. Add onions and garlic back with remaining ingredients. Simmer 1 hour or more.

This is a mild sauce. Increase last three ingredients to make more spicy sauce. Serve over spaghetti. Suggestion: Boil spaghetti rapidly 10 to 15 minutes. Do not cover. Pour into a colander and rinse with hot water to get starch out. Makes about 2 quarts.

Delicious Meatless Enchiladas

Sauce:

- 2 c. tomato sauce
- 2 c. water
- 2 tbsp. dried onion soup
- 1 tsp. salt
- 1 tsp. garlic salt
- 1 tsp. oregano
- 2 bouillon cubes, mashed, or other beef flavor

Mix and heat. If too acid add 1/2 tsp. sugar and pinch of soda.

Filling:

- 2 c. ground gluten browned in
- 2 tbsp. cooking oil
- 2 c. Monterey cheese, grated
- 1 onion, finely chopped
- 2 green onions, finely chopped
- 8 tortillas (Recipe in *Starch* Section)

Fry tortilla lightly in 1/4 c. hot oil. Remove and dip both sides in sauce. Spread with gluten, a little cheese and onions. Roll up and place in baking dish. Continue with all tortillas. Top with remaining sauce. Sprinkle with remaining grated cheese and onions or with Parmesan grated cheese. Bake 350 degrees 20 minutes. Serves 4-8.

Tempting Wheat Salad

- 3/4 c. sifted cracked wheat
- 2 c. water
- 1/4 tsp. salt
- 1 tsp. butter or margarine
- 1/2 c. tomato soup sauce from *sauce* section
- 1/2 c. mayonnaise or salad dressing
- 1 c. ground gluten
- 1/4 c. green pepper, chopped
- 2 tbsp. green onion, chopped
- 3/4 c. celery chopped
- 1-1 1/2 c. tuna flaked

Combine wheat, water, salt and butter in saucepan and simmer covered about 20 minutes. Cool slightly and add remaining ingredients. Chill several hours or overnight for better flavor.

Nutritious Pet Food

2 c. ground gluten
2 tbsp. pwd. eggs
2 tbsp. pwd. milk
1 tbsp. beef flavor

Combine all ingredients. Spread on cookie sheet and bake 300 degrees 20-25 minutes or until well dried out. Store on shelf in closed container and serve as is to pets. Suggestions: When serving cats, pour over liquid from any canned fish. When serving dogs, mix in leftover meat gravy, broth, T.V.P. or canned dog food.

Chapter 10

GLUTEN AS PART MEAT

Gluten as a Meat Extender

Gluten can be used as a meat extender and can be adapted to any of your favorite recipes by using 1 cup ground gluten and 1 cup or 1/2 pound ground round for each pound of meat needed. If meat is all lean, add 1 or 2 tablespoons olive oil or cooking oil. With chicken or tuna use equal parts also.

If you desire to further extend the meat, use 3 or 4 parts gluten to one part meat, chicken or tuna. Add extra flavor by using seasoning as if you were using gluten in place of meat. See *Using Gluten in Place of Meat* for suggestions.

Recipes

Versatile Meat Balls or Patties

2 c. ground gluten
1 lb. ground round
1 can chicken with rice soup
4 eggs, beaten
2 c. bread crumbs, finely ground

$1 1/2$ tsp. salt
1/2 tsp. pepper
1/2 tsp. poultry seasoning
1 med. onion, finely chopped

Mix all ingredients together. Form into about 75 one-inch balls. Bake on greased baking sheet 350 degrees 20 minutes, or fry in small amount of oil. These can be stored in freezer. Take out the amount you need for a meal and keep the rest frozen. Suggestion: Use as regular meat balls. Use any of the sauces in the sauce section. Make about 20 good size patties and cook on a grill, or add 1/2 teaspoon of liquid smoke and fry in 2 tablespoons of oil, or bake in 350 degrees about 40 minutes. Serve as a hamburger with favorite trimmings. May also be used in a foil dinner, base for sausage franks or poultry dressing.

Special Franks

1/2 Recipe Meat Balls (see above recipe)	1/4 tsp. garlic powder
1 lb. pork sausage	2 wieners, ground

Mix together thoroughly. Form in wiener-size links. Bake 350 degrees 30 minutes. Serve as you would a hot dog with all the trimmings, as a corn dog and in casseroles.

Franks in Spicy Rice

4 franks (see above recipe)	1/2 tsp. chili powder
2 tbsp. bacon drippings or oil	1/4 tsp. cloves
1 clove garlic, finely chopped	3 c. rice, cooked
2 tsp. Worcestershire sauce	2 c. tomatoes, canned
1/8 tsp. cayenne pepper (optional)	1 1/2 tsp. salt
	Dash of pepper

Cut franks in 1 inch pieces and saute with garlic in oil. Combine in baking dish with remaining ingredients. Sprinkle 1/2 cup grated American cheese on top. Bake 400 degrees 25 minutes. Serve hot. Serves 6-8.

Pork Sausage Extender

1 c. ground gluten	1/8 tsp. cayenne pepper or more
1/2 lb. sausage, bulk	
1/2 tsp. sage or more	1/4 tsp. salt

Mix together thoroughly and use as regular sausage.

Meat Loaf With Applesauce

1 lb. ground round	1 c. applesauce
1 recipe Pork Sausage Extender	1/2 c. onion, minced
	1/2 c. celery, chopped fine
1 egg, beaten	2 tbsp. flour
1 c. one-minute oats	Dash nutmeg

Mix together all ingredients, pack in 5 x 9 inch loaf pan. Bake in a 350° oven about one hour. Serves 10.

Sausage Skillet Dish

1 recipe Pork Sausage Extender
2 tbsp. minced onion
1 c. minute rice Extender
1 can tomato soup
2 tbsp. catsup

In fry pan crumble sausage, brown with onion. Add remaining ingredients. Cover and cook slowly about 30 minutes. Serves 5-6.

Hawaii Chicken With Meat Balls

2 chicken breasts, split in half
1/4 c. flour
1/2 tsp. salt
Shake of pepper
1/3 c. cooking oil
24 "meatless balls," unbaked
1 recipe sweet and sour sauce

Coat chicken in mixture of flour, salt and pepper. Heat oil in fry pan and brown chicken on both sides. Arrange chicken and meatless balls in baking dish. Make recipe of sauce but do not add pineapple or green pepper. Pour sauce over chicken and balls, bake 350 degrees 30 minutes. Place pineapple bits or slices and green pepper cut in strips around chicken and balls, bake 20 minutes or until chicken is tender. Serve with fluffy steamed rice. Serves 4.

2-Tasty Meatloaves in One Mix

2 c. ground gluten
1 lb. ground round
$1^1/_2$ c. mild cheese, diced
2 eggs, beaten
1 lg. onion, chopped
1 lg. green pepper, chopped
2 tsp. salt
1 tsp. pepper
1 tsp. celery salt
1/2 tsp. paprika
2 c. milk
2 c. dry bread crumbs

Combine all ingredients thoroughly. Press one-half of mix in greased loaf pan. Freeze to use later or bake 350 degrees 1 to $1^1/_2$ hours. Serves 5-6.

Form loaf of remaining mixture. Brown on all sides in a heavy pan with a lid. Add 1 cup water or tomato juice or beef stock and 3 or 4 cups of raw vegetables, such as potatoes, carrots, onions, etc. Simmer 1 hour. Use juice for good rich gravy base. Serves 8-10.

Meat Loaf With A Special Sauce

2 c. ground gluten
1 lb. ground round
1 c. bread crumbs
1 med. onion, finely chopped
1 1/2 tsp. salt
1/2 tsp. pepper
1/2 8-oz. can tomato sauce

Mix all ingredients thoroughly and press into a deep loaf pan. Pour sauce over top.

Sauce

1/2 8-oz. can tomato sauce
2 tbsp. prepared mustard
1 c. water
1/4 tsp. nutmeg
2 tbsp. brown sugar
2 tbsp. vinegar

Combine all ingredients and pour over loaf (allow 1" at top for sauce to bubble). Bake 350 degrees 1½ hours. Remove loaf to platter. Slice and pour sauce over each slice. Serves 10-12.

Meat Loaf and Gravy

1 lb. ground round
2 c. ground gluten
1 pkg. dry onion soup
1 tsp. Italian Seasoning, optional
1 can cream of mushroom soup
1/3 can milk

Combine meat, gluten and onion soup. Pack in deep baking pan or loaf pan. Combine soup and milk, pour over loaf, cover tightly with foil. Bake at 350° for 1 hour 15 minutes. As this loaf bakes, meat juices and soup combine for a tasty gravy. Serve over mashed potatoes, noodles, rice or toast. Serves 8-10.

Ground Round Soup

1 c. ground gluten
1/2 lb. ground round
1 small onion, chopped
2 sticks celery, chopped
1½ tsp. salt
1 tsp. pepper

3 c. water
3 med. carrots, grated
3 med. potatoes, grated
1 tsp. A-1 sauce or steak sauce
1 can tomato soup

In large saucepan crumble gluten and meat in water. Add remaining ingredients and cook until vegetables are tender. (Do not overcook.) Serves 5-6.

Quick Meat Pie

2 c. ground gluten
1 onion, chopped
4 tbsp. olive oil or cooking oil
1 can chili without beans
2 c. grated cheddar cheese
3 small pkgs. Fritos
2 tsp. Parmesan cheese

Brown gluten and onion in oil. Add remaining ingredients, saving a few Fritos to put on top. Pour into buttered casserole in which you have sprinkled Parmesan cheese or number 1 seasoned topping from Tasty Toppings. Serves 5-6.

Enchiladas Casserole

16 oz. pkg. corn chips
1/2 c. shredded cheese
1 can chili without beans
2 c. ground gluten
1-15 oz. can enchilada sauce
18 oz. can tomato sauce
1 tbsp. instant onion
1 c. sour cream (optional)

Reserve 1/2 cup of corn chips. Combine in saucepan remaining chips, 1 cup of cheese, chili, gluten, sauces and onions. Bring to boil and stir to melt cheese. Put in baking dish 11 x 7 x $1^{1}/_{2}$. Bake 350 degrees uncovered, with sour cream spread on top and sprinkled with 1/2 Cup cheese until cheese melts. Put reserve corn chips around edges. Bake 5 minutes more. Serves 6-8.

Easy Skillet Chili

1 c. ground gluten
1/2 lb. ground round
1/2 c. celery, chopped
1 lg. can pork and beans
11-lb. 12 oz. can tomatoes

1 lg. onion, chopped
2 tbsp. cooking oil
1/2-1 tsp. chili powder
1 c. chili sauce

In large skillet brown gluten, meat, onion and celery. Drain. Puree tomatoes in blender and add with remaining ingredients to meat mixture. Simmer 15 to 20 minutes. Stir several times. Serves 10-12.

Well-Seasoned Noodle Casserole

1 c. ground gluten
1/2 lb. ground round
2 tbsp. cooking oil
1 c. frozen mixed vegetables
  or frozen peas & carrots
  or frozen peas
1 tsp. salt
1/4 tsp. pepper
2 onions, sliced
1 c. noodles
1 can tomato soup
3/4 c. seasoned topping
  from topping section

Cook noodles in 1 quart boiling water 10 minutes. Drain. Brown meat, gluten and onions in oil. Add noodles, vegetables, salt and pepper. Pour into a buttered casserole. Pour tomato soup over top and sprinkle with seasoned topping of your choice. Bake in 350 degree oven 30 minutes. Serves 5-6.

Super Supper In-A-Dish

1 c. ground gluten
1/2 lb. ground round
1/3 c. green pepper, chopped
1/3 c. onion, chopped
2 tbsp. rice, uncooked
1/2 tsp. pepper
$1\frac{1}{2}$ tsp. salt
1 c. carrots, thinly sliced
1 can tomato soup
  mixed with
1 c. water

Combine ground round and gluten and pat evenly in bottom of greased baking dish. Add remaining ingredients as listed. Bake 350 degrees 1 hour covered, 15 minutes without cover. When serving, make sure to dip each serving all the way to the bottom of dish. Serves 6-8.

Quick Luncheon Dish

1 c. ground gluten
1/2 lb. ground meat
1/3 can milk
1 tbsp. dry onion soup

1 can cream of mushroom soup
1 tsp. dry parsley flakes

Brown meat and gluten. Combine remaining ingredients. Simmer 20 minutes. Serve over rice or buttered toast. Serves 6-8.

Ground Round Special

1 c. ground gluten
1/2 lb. ground round
2 tbsp. cooking oil
1 can spinach, drained (or 2 bunches fresh cooked & drained)
4 eggs, slightly beaten
1 tsp. salt
1/2 tsp. pepper
1/8 tsp. garlic salt

Brown gluten and meat in oil. Add onion and cook until tender. Drain. Add remaining ingredients. Toss gently with spatula just until eggs are done. Serve hot. Serves 6-8.

Chicken Layered Casserole

3/4 c. ground gluten
1 c. diced, cooked chicken
4 tbsp. butter
4 tbsp. flour
1 c. chicken broth
1 lg. can evaporated milk
1/2 c. water
1 tsp. salt
3 c. rice, cooked
1/4 c. pimento, chopped
1/3 c. green pepper, chopped

Melt butter, add flour and blend. Add broth, milk and water and cook until thick, stirring constantly. Add salt. Mix gluten and chicken. Alternate layers of rice, chicken-gluten mix, vegetables, and sauce, in a greased 10 x 6 x $1\frac{1}{2}$ inch baking dish. Pour remaining sauce on top. Bake in 350° oven for 30 minutes. Serves 8-10.

Turkey or Chicken Salad

1 tbsp. lemon juice
2 c. apples, diced
3-4 c. chicken, chopped
1/2 c. ground gluten
1 c. salad dressing
1 tsp. salt
1/4 tsp. pepper
$1\frac{1}{2}$ c. celery, chopped
$1\frac{1}{2}$ c. grapes, green seedless

Place apples in large bowl, sprinkle with lemon juice. Add remaining ingredients and toss lightly. Chill. Serves 15.

Tuna-Chicken Bake

1 c. ground gluten
1 c. tuna, flaked
1 c. macaroni
1/2 c. green pepper, chopped
3 tbsp. oil
2 tbsp. flour
$1\frac{1}{4}$ c. milk
1 can cream of mushroom soup
1/4 c. pimento, chopped
2 tbsp. onion, chopped

Cook macaroni 7 minutes. Drain and set aside. Cook onions and green pepper in hot oil until tender. Blend in flour. Stir in milk and cook until thick. Add soup, macaroni and remaining ingredients. Pour into greased $1\frac{1}{2}$ qt. casserole. Bake at 350 degrees for 30 minutes. Suggestion: Garnish with sliced almonds. Serves 6-8.

Quick Make-Ahead Tuna Casserole

1 c. ground gluten
1 c. tuna, flaked
1 c. elbow macaroni, uncooked
2 cans cream of mushroom soup
1 med. onion, chopped
1 c. cheddar cheese, grated

Combine all ingredients. Press in a 2-quart casserole pan. Let set in refrigerator 5 to 6 hours or overnight. Stir and bake at 350 degrees for 1 hour covered, and 15 minutes uncovered. Serves 6-8.

New Way Tuna Souffles

8 slices bread, cubed
1 c. mild cheese, grated
1/2 c. ground gluten
1/2 c. tuna, flaked
3 eggs, beaten
2 c. milk
1/2 tsp. salt
1/4 tsp. pepper
1/4 tsp. paprika
1/2 tsp. prepared mustard

Place one-half of bread cubes in the bottom of greased medium size loaf pan. Spread one-half of cheese over bread. Sprinkle with gluten and tuna. Add remaining bread and top with rest of cheese. Combine rest of ingredients and pour over mixture. Allow to sit 15 minutes to soak in bread. Bake at 350° for 45 minutes. Serve immediately. Serves 5-6.

Spanish Dinner

1 c. ground gluten
1/2 lb. ground round
2 tbsp. oil
1/4 lb. cheese, grated
1/2 c. rice
1 beef bouillon
1 med. pkg Frito corn chips, crumbled

1 bottle taco sauce
1 head lettuce
2 lg. tomatoes
1 avocado
1 cucumber
6 radishes

Brown gluten and ground round in oil. Add cheese and stir till melted, cover and turn off heat. Cook rice in one cup of water with bouillon, then add meat mixture. Make a tossed salad of vegetables. To serve, place portions of meat and rice mixture on each plate with a portion of salad on top. Sprinkle with crumbled Fritos and taco sauce to your taste. Serves 5-6.

Meal-In-One Salad

13 oz. lemon Jell-o
1/2 c. hot water
1/2 c. ground gluten
1/2 c. tuna, flaked
1 can chicken gumbo soup
2 tbsp. green pepper, chopped

2 tbsp. onion, chopped
2 tbsp. celery, chopped
2 tbsp. cucumber, chopped
1/2 c. cream, whipped
1/2 c. mayonnaise

Dissolve Jell-o in hot water. Set half way. Blend whipped cream and mayonnaise with remaining ingredients. Fold into Jell-o. Set until firm. Cut into squares and serve on a bed of shredded lettuce leaves. Serves 5-6.

Pizza Burger

1 c. ground gluten
1/2 lb. ground round
1/2 onion, chopped
2 tbsp. olive oil
2 tbsp. molasses

Salt and pepper to taste
1-10 oz. can pizza sauce with cheese
6 hamburger buns
12 cheese slices of your favorite cheese

1 tsp. dry mustard
1 tsp. chili pwd.

Lightly brown gluten, meat and onions in oil. Add molasses, mustard, chili powder, salt, and pepper. Simmer uncovered 10 minutes. Cut buns in half, top each with a cheese slice. Broil 3 minutes or until cheese is bubbly. Spoon meat mixture over each bun. Makes 1 dozen individual burgers.

Sour Cherry Pilaf

1/2 lb. ground round
1 c. ground gluten
1 lg. onion, chopped
2 tbsp. oil
2 c. sour cherries

$1\frac{1}{4}$ c. sugar
1/2 tsp. cinnamon
2 tsp. dill seed
Tiny pinch saffron, optional

Brown meat, gluten and onions in oil, drain. Heat cherries and sugar until sugar is melted, add remaining ingredients, pour over meat mixture. Stir well, simmer 15 minutes, and serve over rice. Serves 6-8.

Iranian Koresh

1/2 lb. ground round
1 c. ground gluten
2 tbsp. oil
2 cans tomato sauce
1 tbsp. cinnamon
1/2 tsp. Worcestershire sauce
1 tsp. salt

1/4 tsp. pepper
1/2 c. water
1 onion, diced
2 or 3 carrots, sliced
1/2 c. celery, chopped
1/2 c. peas

In large fry pan brown meat and gluten in oil and drain. Mix in remaining ingredients and simmer until vegetables are done but still crisp. Serve over hot rice.

Canned Tomatoes Layered Casserole

2-3 carrots, sliced
2-3 potatoes, cubed
Salt and pepper
1/2 lb. ground round

1 c. ground gluten
1 c. onion, chopped
2-3 c. canned tomatoes
   blend in blender

In deep baking dish place layers of ingredients in order given. Pour blended tomatoes over all ingredients to cover. Bake 375 degrees about 1 1/2 hours. Makes 6-8 servings.

Barbecue Hamburger

1 lg. onion, chopped
2 tbsp. oil
1 lb. ground round
2 c. ground gluten
1 1/2 tsp. salt
1/4 tsp. pepper
1 can chicken gumbo soup
1 tsp. steak sauce
1/2 c. barbecue sauce
1 tsp. chili powder
1 clove garlic, minced

In fry pan brown onion in oil, add meat and gluten and slightly brown. Mix in remaining ingredients, simmer 30 minutes, serve on hamburger buns. Filling for 2 doz. hamburgers.

Chili Balls

1 c. ground gluten
1/2 lb. ground round
1/2 c. rice, uncooked
1 sm. onion, chopped
1 tsp. salt
1/4 tsp. pepper
2 tbsp. olive oil
2 cans tomato soup
1 can water
1/2 tsp. chili powder or more
1/2 tsp. ground cumin

Mix first 6 ingredients. Shape into 16 to 20 balls. Brown lightly on all sides in hot oil or bake at 350 degrees for 30 minutes. Combine soup, water, chili powder and cumin. Add to balls. Suggestion: Serve over 1 cup rice steamed 20 minutes in 2 cups water, 1 teaspoon each salt, dry parsley flakes, and butter. Serves 5-6.

Turkey or Chicken Stuffing

Dressing

3 c. ground gluten
2 c. bread cubes
2 c. cornbread, crumbled
1/4 c. celery, chopped
1/4 c. onion, chopped
1/2 tsp. thyme
1/2 tsp. sage
1/4 c. butter
2 eggs, beaten
1 tsp. salt
Chicken or turkey broth

Combine all ingredients. Moisten with broth using milk if more liquid is needed. Use as stuffing in turkey or chicken or bake in pan at 350 degrees for 30 minutes. Serve with Cranberry Sauce from sauce section.

Broth

Giblets from turkey or chicken
5 c. water
1 tbsp. parsley, chopped
1 bay leaf
1 onion, chopped
4 cloves
1 tsp. salt
1 tsp. pepper

Combine all ingredients. Cook until giblets are tender. Drain and pour over dressing. Serves 5-6.

Chicken Bake Supreme

1/2 c. ground gluten
1/2 c. rice, uncooked
1 can cream of chicken soup
$1\frac{1}{4}$ c. water
1 pkg. onion soup
1 fryer, cut up

Combine all ingredients except chicken in a 2 quart baking dish. Salt and pepper each chicken piece and place on top. Bake uncovered at 350 degrees for $1\frac{1}{2}$ hours or until chicken is tender. Serves 6-7.

Chapter 11

OTHER GLUTEN RECIPES

Mock Steaks

To prepare gluten for mock steaks refer to *Baking and Grinding Gluten for Quick Use,* page 48. Slice baked gluten into 1/ inch thick slices. Two recipes are given below for seasoning an cooking it to give it a taste and texture very similar to steaks.

Chicken Fried Steak

6-8 slices baked gluten
1 can Campbell's
   Consomme beef soup
1/2 can water
1 cup bread crumbs,
   finely ground
2 egg yolks, beaten with
1 tbsp. milk and
1 tsp. A-1 sauce
2 egg whites, beaten
   until frothy
Olive oil or cooking oil

In a one and one-half quart saucepan combine soup and water and add gluten slices. It is important the slices are completely covered with broth. Simmer one to one and one-half hours, or until most of broth is gone. Twice during simmering move steaks from bottom of pan to top so all will be seasoned well. Remove steaks and drain well by pressing each steak firmly between 2 saucers. As you press the saucers together, tip them on edge so the broth will run out. Blot each steak with a paper towel. The steaks will have a better texture if you remove most of the broth.

Dip each steak in beaten egg yolk mixture, then in bread crumbs, then in egg whites making sure to dampen all the crumbs with the whites, then dip in crumbs again. This procedure can be done early in the day.

Pour 1/2 inch of oil into the fry pan and heat on medium heat Brown the steaks to a golden brown on each side and then drain them on a paper towel. Serve hot, as a steak, with catsup or steak sauce. Suggestions: Serve steaks with your favorite gravy or a sauce from the sauce section. Leftover steaks may be served on toasted buns as a sandwich. They may also be ground through a food chopper and added to pickle relish and salad dressing for a sandwich spread.

Breaded Veal Steaks

6-8 slices gluten
3 c. hot water
3 chicken bouillon cubes
3/4 c. bread crumbs, finely ground
1 tsp. poultry seasoning
1/2 tsp. salt
2 eggs, beaten
Olive oil or cooking oil

In a one and one-half quart saucepan dissolve bouillon cubes in water and add gluten slices. It is important that the slices be covered with broth. Simmer one to one and one-half hours, or until the broth is almost gone. Twice during simmering, carefully move steaks on the bottom of the pan to the top so all will be seasoned well. Remove steaks and drain them well by pressing each steak firmly between 2 saucers. As you press the saucers together, tip the saucers on edge so the broth will run out. Blot each steak with a paper towel. The steaks will have a better texture if you remove most of the broth.

Mix seasonings in bread crumbs. Dip each steak in eggs and then in crumbs. Pour 1/2 inch oil in fry pan and heat on medium heat. Brown steaks well on each side. Serve hot.

Great Ideas for Lunchmeat

To prepare gluten as lunch meat refer to *Baking and Grinding Gluten for Quick Use,* page 48. Slice the baked gluten to 1/8 inch thick. Two recipes are given for very tasty lunch meat.

Spicy Lunchmeat

6 very thin slices baked gluten
1 tsp. dill seed
1/4 tsp. marjoram
1/4 tsp. tarragon
1 tbsp. Schilling Beef Flavor base
2 c. water

In a small saucepan simmer sliced gluten in water with seed and seasonings until most of water is gone, about 1 hour. Remove slices and drain well by pressing each slice firmly between two saucers. As you press, tip the saucers on edge so broth will run out. Blot each slice very well with paper towel. Chill and serve as regular lunch meat or store in refrigerator 4 or 5 days in closed container. *Suggestions:* Put lunch meat through food chopper with a hard boiled egg and sweet or mustard pickled relish, add mayonnaise and use as sandwich spread.

Sandwich Spread

6 thin slices baked gluten
2 c. water
1 tbsp. ham flavor
(Bernard ham flavor soup seasons well, or use 1/4 e. T.V.P. ham bits, or omit ham seasoning and simmering and add 1/2 e. chopped ham as you put slices, onion soup? relish and egg through food chopper.)

1 tbsp. dry onion soup
2 tbsp. sweet pickle relish
1 egg, hard boiled

In a small saucepan simmer sliced gluten in water, ham flavor and onion soup until most of water is gone, about 1 hour. Remove slices and drain well by pressing each slice firmly between two saucers. As you press, tip the saucers on edge so broth will run out. Put slices through food chopper with relish and egg. Add mayonnaise and mustard to moisten. Chill and serve as sandwich with lettuce and tomatoes or on crackers. *Suggestion:* Add extra mayonnaise and use as dip with chips or raw vegetables.

Cook a Little Magic with Raw Gluten

To prepare raw gluten refer to *Extracting Gluten in Three Easy Steps,* p. 47, Step I and Step II.

Mini Pizza

1 c. raw gluten
1 tbsp. Schilling Beef Flavor
1/4 tsp. garlic salt

1 tbsp. dry onion soup
1/3 c. Parmesan grated cheese

Roll gluten to 1/8 inch thickness. While rolling out gluten, cut edges 1/4 inch at about 2 inch intervals for easier rolling. Sprinkle remaining ingredients in order given. Roll as cinnamon roll, then cut 1 inch thick slices. Place on baking sheet well greased with shortening. Bake 350 degrees 20 to 25 minutes. Serve hot. Makes 12 to 15 Mini Pizzas.

Mock Swiss Steak With Vegetables

1 c. raw gluten
1/2 c. whole wheat flour
1 tbsp. Schilling Beef Flavor
1 tsp. salt
1/2 tsp. pepper
Olive oil or cooking oil
1 egg, beaten

1 onion, sliced
1 green pepper, sliced
2 c. tomatoes, canned
1/2 c. beef broth or
   Campbell's Consomme Soup
Carrots, cut in 1/2 inch slices
Potatoes. cubed

Roll gluten to 1/4 inch thickness and cut in about 2 inch squares. As gluten is rolled, score or clip edges 1/4 inch at about 2 inch intervals. This makes the gluten roll easily. Dip each steak in egg, then in flour to which you have mixed in beef flavor, salt and pepper. Brown well both sides of steaks in fry pan with 1/2 inch olive oil, cooking oil or shortening. (These may now be stored in refrigerator or freezer for future use if desired.)

In saucepan add remaining ingredients and cook to just done. Pour oil from fry pan, return steaks to bottom of pan, add hot cooked vegetables in layers in this order: onions, pepper, carrots, potatoes. Pour tomatoes and broth over all. Heat through and serve. Do not allow to simmer or boil as steaks will take on a spongy texture. Serves 8.

Treats for Snacking

1-1 1/2 c. raw gluten
1/4 c. dry onion soup

1 tbsp. Parmesan cheese
1 tsp. dry parsley

Mix together all ingredients thoroughly by stretching and pulling about five minutes. Cut in small one inch pieces. Bake 350 degrees about 20 minutes. About 25 treats.

Tasty Wheat Roast

1 1/2 c. raw gluten
1 tbsp. dry onion soup

1/4 c. Parmesan cheese
1 tbsp. parsley flakes

1/4 tsp. garlic salt
1 tsp. beef flavor base
1-2 tbsp. Worcestershire sauce
or steak sauce

Mix together soup, salt, cheese and flakes. Put gluten through food chopper, with medium blade, while adding dry ingredients. With hands, work sauce in and place in well-greased loaf pan. Bake at 350 degrees until done, about 1 hour. Slice thin and use as roast, chip beef, steaks, on pizza, or cube and add to cooked stews. Do not simmer or boil in liquid as the texture will be spongy. 20-30 slices.

Jerky

Miracle Wheat Jerky

1 c. raw gluten
1 can Campbell's
    Consomme Beef Soup
1/2 can water

1/2 tsp. garlic salt
1 tbsp. kitchen bouquet
    (for color)
1 tbsp. liquid smoke

Roll gluten to 1/8 inch thickness (no less). As gluten is rolled, cut edges with knife 1/4 inch at 2 inch intervals, which helps gluten roll out easier. Cut rolled gluten in strips about 6 to 8 inches long, 1 inch wide. In a one and one-half quart saucepan, combine remaining ingredients and bring to boil. Drop cut strips of gluten into hot broth. Lower heat and simmer 45 minutes to 1 hour or until most of broth is gone. Gently stir 2 or 3 times during simmering so all strips will be seasoned well. Remove strips from broth and drain well in strainer. Put strips in saucepan, add 2 tablespoons more of liquid smoke, and mix well. Place strips so as not to touch each other on a square of 1/4 to 1/2 inch hardware cloth or on a similar rack. (Size that will fit in oven.)

Dry out in a 300 degree oven, with door cracked two inches, about 30 minutes. Test for your desired doneness. Dry longer if you feel necessary. Remove from oven, cool and store in closed container. *Suggestions:* Different flavored jerky may be obtained by using chicken, bacon or ham flavor, or use 1 part ham and 1 part beef flavor. Also 2 or 3 tablespoons honey may be added. To make it hot, add hot sauce or cayenne pepper to broth. Makes about 2 dozen 5 inch strips.

Jerky, Ages Ago

The primitive man living in deserts and high in the mountains dried meat for preservation. When weather conditions were not suitable, they learned to dry by fire. The primitive man developed a taste for the wood smoke. Drying became a natural preservative.

Whatever meat could not be cut off was left on the carcass for drying. Some believe this was first done in Asia and Africa. Dried meat has been found in the lower Egyptian Nile and in Northeastern Mongolia. It could be carried easily by travelers on foot or horseback as it was light in weight and it stored well.

As warriors and travelers journeyed along the Mediterranean Sea, wild food animals became scarce. The few farmers that lived along the way were unable to raise enough livestock to meet the needs of the villagers. Market centers were opened. Tribesmen, hunters and farmers brought meat for sale. Jerky became the main product.

The art of drying meat came to our country through migrations across the Bering Straight. No one can fix a date except it was many centuries ago. The Portuguese explorers called the dry meat Xarque, the Spanish, Char-qui, the Mexicans, Tasajo and the English, Jerky. The word Jerky is now used mostly in the United States. When Europeans first came to America, they found most of the Indians using crude stone knives for cutting their meats. A few tribes had developed lead and copper knives. The meat of many animals was used: bison, deer, moose, elk, bear and caribou. Hunters trapped animals during the summer when the meat was at its best. What they couldn't use was dried.

The meat was hung and dried by sun, wind and smoke, Sometimes it would be necessary to pound it in small bits because it was so hard to chew. The drying develops the flavor from natural enzymes in the meat. The smoke lends flavor to the outside.

Trappers and explorers have enjoyed Jerky for many years. Today, scouts can be found with Jerky on a hike along trails or sitting around campfires.

The author sincerely hopes you will try and enjoy her recipe for "Miracle Wheat Jerky."

Section Three
USING BRAN

Bran Flakes

Canned Old Fruit Cake

Quick Pancakes

Raised Whole Wheat Pizza Dough

Cookies

Jar Showing Bran, Starch, and Water

Chapter 12

USING BRAN AND STARCH WATER

Bran is Beneficial

Bran is worth saving and worth using! To gain full food value of the whole wheat kernel we need to use the bran too. Bran is a natural source of trace minerals which are essential in the balance of the body. Bran provides roughage in the diet that aids in the muscular action of the digestive system. Thus the bran helps normal elimination.

Bran can successfully be used in banana bread, oatmeal and applesauce cake by adding 2 tablespoons to each loaf. It makes a light texture. Note its many uses in the following recipes. *It is important that the whole wheat flour used for all recipes in this section be freshly ground or sifted.*

Cereal Flakes

> 1 c. bran
> 1 tsp. baking powder
> 1/2 tsp. salt
> 1 tbsp. sugar
> 2 tbsp. oil

Mix together all ingredients. Batter should be thin so add a little water if it needs thinning. Pour batter on two greased 11 x 15 inch baking sheets and spread very thin. Bake at 350 degrees for 20 minutes. Break in small pieces and serve as breakfast cereal with milk and sugar. Makes $1\frac{1}{2}$ to 2 cups.

Caraway Crackers

> 1 c. bran
> 3 tbsp. oil
> 1/2 tsp. salt
> 1 tsp. baking powder
> 1/4 tsp. onion salt
> 2 tsp. caraway seed

Mix together all ingredients. Batter should be thin so add a little water if it needs thinning. Pour batter on two greased 11 x 15 baking sheets and spread very thin. Bake at 350 degrees for 20 minutes. Break in small crackers and serve with cheese or other cold cuts as desired.

Quick Pancakes

1/4 c. bran	2 tbsp. sugar
1 3/4 c. whole wheat flour	2 1/2 tsp. baking powder
1 c. milk	1/2 tsp. salt
2 eggs	2 tbsp. melted butter

In blender or mixing bowl beat eggs. Add remaining ingredients and beat until mixed. Bake on medium hot griddle. Serve with butter and hot maple syrup.

Whole Wheat Yeast Muffins

1 tbsp. dry yeast	1 egg, beaten
1 c. warm water	1/3 c. bran
1/4 c. sugar or honey	3 c. whole wheat flour
1 tsp. salt	4 tbsp. cooking oil

Dissolve yeast in water. Add bran, sugar, salt and beaten egg. Add flour and oil and beat until smooth. Dough will be slightly stiff. Let raise 20 minutes. Drop by spoonfuls in greased muffin tins and let raise to double. Bake 15 minutes at 400 degrees. Makes 18 medium-size muffins.

Company Brownies

4 eggs	1 1/4 tsp. baking powder
1 1/2 c. whole wheat flour	5-6 tbsp. cocoa
1/4 c. bran	1 tsp. vanilla
1 1/2 c. sugar	1 c. nuts
2/3 c. oil or soft margarine	

Beat eggs till fluffy. Add sugar and beat well. Add remaining ingredients and mix thoroughly. Pour in greased dripper pan. Bake 350 degrees for 25 to 30 minutes. Frost with favorite chocolate or white frosting. Makes 24 to 30 squares.

Fresh Apple Pudding

1 c. sugar	1 tsp. soda
1/4 c. soft butter	1/2 tsp. baking powder
1 egg	1 tsp. cinnamon
2 c. shredded unpeeled apples	1/2 tsp. nutmeg

1 c. whole wheat flour   1/4 tsp. salt
1/4 c. bran   1/2 c. chopped nuts

Cream butter and sugar. Add egg and beat hard. Add apples to creamed mixture. Stir in dry ingredients. Fold in nuts. Bake in 9 x 9 greased pan 360 degrees for 50 minutes. Cut in squares. Serve hot as is or with whipping cream. Makes 9 to 12 servings.

Brown Sugar Squares

2½ c. whole wheat flour   2 c. brown sugar
2½ tsp. baking powder   3 eggs
1/2 tsp. salt   1/4 c. bran
2/3 c. shortening   1 c. nutmeat
   (butter makes it richer)   6 oz. pkg. chocolate chips

Sift flour and dry ingredients and set aside. Cream shortening and sugar very well. Add eggs one at a time beating after each addition. Mix in bran and add dry ingredients. Stir in nuts and chips. Mixture will be thick. Spread on greased 10 x 15 inch pan and bake at 350 degrees for 30 minutes. Cut into squares when almost cool. Makes 5 dozen squares.

Chewy Bars

2 tbsp. butter or margarine   5 tbsp. white or whole
2 eggs, beaten   wheat flour
2 tbsp. bran   1/8 tsp. baking soda
1 c. brown sugar, packed firm 1 c. nuts
1 tsp. pure vanilla

In a 9 inch square pan melt butter slightly and remove from heat. Mix together sugar, flour, soda and nuts and set aside. Mix together eggs, bran and vanilla and combine with flour mixture. Pour all over butter but do not stir. Bake 350 degrees 20-25 minutes. Turn bottom side up on rack and sprinkle with powdered sugar. Cut in bars. Makes 9 large squares.

Delicious Fudge Nut Bars

4 eggs   2 cups margarine, melted with
1/4 tsp. salt   4 tbsp. cocoa
2 c. sugar

2 tsp. bran   1½ c. whole wheat flour
                1 c. nuts, large pieces

Combine eggs, salt, sugar and bran. Beat until thick. Add margarine, cocoa, mix alternately with flour. Beat well, then add nuts. Pour in greased 8 x 12 pan. Bake at 350 degrees until almost done, about 30 minutes. Sprinkle generously with miniature marshmallows and cover partially with thin chocolate frosting made from sifted cocoa and powdered sugar, margarine and hot water. Makes 2 doz. bars.

Banana-Bran Bread

1/2 c. margarine, soft   2 c. flour
1 c. sugar   1 tsp. soda
2 eggs   1/2 tsp. baking powder
1/4 c. bran   1 c. nuts, chopped
3 bananas (very ripe) mashed

Cream first four ingredients, add mashed bananas and beat well. Stir in flour, soda and baking powder. Add nuts. Pour in 2 medium size loaf which have been greased with shortening. Bake 350 degrees about 1 hour. Makes 2 medium loaves.

Good Whole Wheat Danish Puffs

2 cubes margarine   1 c. boiling water
2 c. whole wheat flour   1 tsp. almond extract
3 tbsp. bran   3 eggs
2-3 tbsp. cold water

Blend 1 cube margarine in 1 cup flour. Add 1 tbsp. bran and cold water. Mix well and press in 9 x 13 ungreased baking pan. Cover and set aside. In sauce pan melt 1 cube margarine in boiling water. Remove from heat and blend in 1 cup flour and 2 tablespoons bran. Add eggs and extract beat well. Pour over pressed dough and bake at 425 degrees for 20-30 minutes or until light brown. Cool 15 minutes and frost while still warm. Cut in squares.

Frosting for Good Whole Wheat Danish Puffs

2 c. sifted powdered sugar   4 tbsp. light cream
1 tbsp. margarine   1 tsp. vanilla

Cream together all ingredients and spread on puffs.

Raised Whole Wheat Pizza Dough

2 tbsp. bran	1 tsp. baking powder
2 c. whole wheat flour	1 tbsp. yeast
1 tsp. salt	3 tbsp. shortening
2 tsp. sugar	1/2 c. warm water

Dissolve yeast in water and sugar. Mix in remaining ingredients. Knead two minutes. Divide dough and roll each in a 12 inch circle and place on pizza pan. Let rise 15-20 minutes. Spread with favorite filling. Bake at 425 degrees about 25 minutes. Makes 2–12" pizzas.

Bran Batter for Wheat Balls

1 c. whole wheat flour	2 tbsp. cooking oil
2 tbsp. bran	Oil for frying
1/2 tsp. salt	1 recipe Wheat Balls,
1 egg	Using Gluten in place of Meat
3/4 c. milk	

Mix dry ingredients. Add egg, milk, 2 tbsp. oil and beat until smooth. Dip each baked wheat ball in batter, fry 2-3 minutes in hot deep oil. Suggestion: Place electric deep fat fry pan in center of table. Using fondue forks let each one cook his own. For added flavor serve with sauce from sauce section. Batter 25-30 Wheat Balls.

The Crown of Cornbread

1 c. cornmeal, freshly ground if possible	1/2 tsp. salt
3/4 c. whole wheat flour, freshly ground or sifted	1/4 c. sugar
2 tbsp. bran	1 egg
4 tbsp. baking powder	1 c. milk
	1/4 c. shortening

Combine all ingredients and beat well. Pour batter in greased 9 inch square pan. Bake 425 degrees 20-25 minutes. Makes 9 servings.

Cream Puffs

1 c. water	4 eggs
1/2 c. butter or margarine	2 tbsp. bran
1 c. (scant) whole wheat flour	

Heat water and butter to rolling boil, stir in flour and bran over low heat until mixture leaves pan and forms a ball. Takes about a minute. Remove from heat. Beat eggs in flour mixture one at a time, until smooth and velvety. Drop from spoon onto greased baking sheet. Bake 45 to 50 minutes at 400 degrees or until dry. Cool slowly. Makes 8 large puffs. Remove inside portion with spoon and use for chocolate eclairs, or fill with whipped cream and fresh fruit, or fill with custards, ice cream, salads such as chicken or tuna. Makes 8 large puffs.

Canned Old Fruit Cake

1 qt. old canned fruit-peaches, apricots, pears or plums	1 tsp. baking powder
4 tsp. soda	1 tsp. cinnamon
1/4 c. bran	1 tsp. nutmeg
2 eggs	2 c. sugar
$4^1/_2$ c. whole wheat flour, sifted*	1 c. cooking oil
1 tsp. salt	1 tsp. vanilla

*Amount of flour will vary according to amount of juice in fruit. Make medium thick batter.

Blend fruit in blender, pour in mixing bowl. Stir in soda (it will foam). Beat in eggs and bran. Sift dry ingredients together and add alternately with oil and vanilla. *Beat thoroughly.* Pour in greased 9 x 12 pan or 2 loaf pans. Bake at 350 degrees for 50 to 60 minutes or until done to touch.

Starch Water

Be Thrifty: Use Starch Water

The starch saved in Step III in *Extracting Gluten in Three Easy Steps,* page 48, can be used to good advantage. It comes from the endosperm or bulk of the kernel of wheat. Use the wheat nutrients which it holds in the following recipes:

Starch for Thickening

To thicken gravy, use 1 to 2 tablespoons very thick starch to each cup of liquid. To thicken puddings, use 2 to 3 tablespoons very thick starch to each cup of liquid.

Good Brown Gravy

2 c. starch water
1/2 tsp. onion salt
1/4 tsp. garlic salt
1/8 tsp. black pepper
1/2 tsp. Worcestershire sauce
1 tbsp. Schilling Beef Flavor base
fi-1 tsp. Kitchen Bouquet

Mix together all ingredients and boil until thick.

Handy Crackers

3/4 c. starch water
1/2 c. cornmeal
1/2 c. Fritos, crushed
2 tbsp. cooking oil
1 tsp. baking powder
1/2 tsp. salt

Mix together all ingredients except salt. Batter must be thin. Use water to thin if necessary. Pour batter very thin on two large greased cookie sheets (Teflon pan preferred). Sprinkle with salt and bake 350 degrees for 15 minutes. Loosen edges and carefully turn whole piece over and bake 5 minutes or until slightly browned. Suggestion: Omit Fritos and add 1/4 teaspoon chili powder to batter and/or 2 teaspoons Parmesan cheese.

Corn Tortilla

1 c. starch water
1/2 c. cornmeal
2 tbsp. flour
1 egg
1/2 tsp. salt

Blend all ingredients together in blender. Batter must be very thin. Add water to thin if necessary. Heat lightly greased griddle on slightly less than medium heat and pour on 1/4 cup batter. Cook until it looks dry, then turn over and cook. Do not brown. Use in Delicious Meatless Enchiladas or any recipe using tortillas. Makes about 8 tortillas.

*Versatile Meat Balls
with
Sweet and Sour Sauce*

*Dip for
Raw Vegetables*

*3 Jars Seasoned Topping
with Casserole*

*Cracked Wheat
Salad
with
Tomato Soup Sauce*

Section Four
USING SAUCES & TOPPINGS

Family Home Evening Treat

Ready-To-Go-Cereal

*Chicken with
No. 2
Seasoned Topping*

Chapter 13

SAUCES

Special Sauces

These special sauces add so much and cost so little. There is a sauce for every meat ball or pattie dish. When time is short, choose one of these favorites and simmer on top of the stove or bake at 350 degrees for 20 to 25 minutes with fresh or frozen meat balls or patties. They each have their own magic blend and will add rich flavor to your meals. Serve them by themselves or over fluffy rice, mashed potatoes, buttered toast or cracked wheat.

Hawaiian Sweet and Sour Sauce

1 c. pineapple chunks
2 tbsp. cornstarch
1 tbsp. soy sauce
1 tbsp. pimento, chopped
1/4 c. vinegar
1/2 c. brown sugar
2 tbsp. lemon juice
1/2 c. green pepper, chopped

Drain pineapple, add enough water to make 1 cup juice. Blend together juice and cornstarch. Stir in soy sauce, vinegar and brown sugar. Cook until thickened, add green pepper, pimento and pineapple chunks. Simmer with 2 dozen meat balls 20 minutes. About $2\frac{1}{4}$ cups.

Quick Sauce

2 tbsp. brown sugar
1/3 c. vinegar
1 can tomato soup
Salt and pepper

Combine all ingredients and heat to just boiling. Add 2 dozen meat balls or patties and simmer 20 minutes. About 2 cups.

Curry Sauce

2 cloves garlic, minced
2 lg. onions, sliced
2 tbsp. oil
2 tsp. curry powder
1/4 tsp. chili powder
1 bay leaf

1 beef bouillon cube
1 tbsp. cornstarch
1½ c. tomato juice dissolved in
1 c. hot water

Saute garlic and onion in oil. Add remaining ingredients and simmer with meat balls or patties 30 minutes or until it begins to thicken. About 3½ cups.

Mushroom Sauce

2 tbsp. onion, chopped
1 tbsp. butter or margarine
1/2 tsp. paprika
1 can cream of mushroom soup
1/4 c. sour cream
1/4 c. water

Cook onions in butter until just tender. Add remaining ingredients and simmer 5 minutes. Add 2 dozen meat balls and simmer 20 minutes. Serve over fluffy rice. About 2 CUPS.

Bacon-Mushroom Sauce

2 strips bacon, cut in bits
2 tbsp. flour
1 tbsp. sugar
1/4 tsp. salt
2 c. tomato juice
1 c. mushroom pieces
   browned in
2 tbsp. butter
1/4 c. ripe olives. chopped

Cook bacon to brown, stir in flour, sugar and salt. Heat till bubbly, then add juice. Bring to a good boil. Add mushrooms and olives. Simmer with 2 dozen meat balls or patties 20 to 25 minutes. About 3½ cups.

Rich Flavor Sauce

1/4 c. vinegar
1/2 c. water
2 tbsp. brown sugar
1 tbsp. prepared mustard
1/2 tsp. pepper
1 tsp. salt
1/8-1/4 tsp. cayenne pepper
3 thin lemon slices
1 med. onion, chopped
4 tbsp. butter
1/2 c. catsup
2 tbsp. Worcestershire sauce
1/2-1 tsp. liquid smoke

Simmer first 10 ingredients 5 minutes uncovered. Add catsup, Worcestershire sauce and liquid smoke. Bring to boil, add 2 dozen meat balls and simmer 20 minutes. About 2 cups.

Mustard Sauce

2 tsp. sugar
1 1/2 tbsp. prepared mustard
1 tbsp. lemon juice
2 tbsp. water
1/4 tsp. salt

1 med. egg yolk, beaten
2 tsp. butter
2 tsp. horseradish
1/4 c. cream, whipped

Mix well first 6 ingredients. Cook in double boiler until thick. Stir in butter and horseradish. Fold in cream. Serve over hot meat balls or meat loaf. About 3/4 cup.

Easy Mild Sauce

1 can cream of mush room soup
1 can cream of chicken soup
1 can milk

Mix together and heat. Add desired amount of meat balls and simmer 20 minutes. About 4 1/2 cups.

Tomato Soup Sauce

1 can Campbell's tomato soup
3/4 c. salad oil
3/4 c. vinegar
3/4 tsp. garlic salt
3/4 tsp. onion salt

1/2 tsp. black pepper
4 tsp. sugar
1/4 tsp. paprika
1 tsp. Worcestershire sauce
1/2 tsp. dry parsley flakes

Blend together all ingredients in blender. Chill. Use as dip for *Wheat Balls,* in *Nutritious Wheat Salad from Using Gluten in Place of Meat* section or as dressing for tossed green salad. About 3 1/4 cups.

Cranberry Sauce

2 c. whole cranberry sauce
2 tbsp. water
1/4 c. raisins
1/4 tsp. prepared mustard
Pinch of salt
1/4 c. nuts, chopped

Combine all ingredients in saucepan and heat. Serve over baked chicken and/or versatile meatless balls. About $2^1/_2$ cups.

Chapter 14

TOPPINGS

Tasty Toppings

These tasty nutritious toppings are for year-round seasoning. They are simple and easy to make and can be stored in a spice corner in closed, labeled containers for handy special occasions and everyday use. Important: Toppings must be thoroughly dried if stored on the shelf or they will mold. Two types of toppings are given, Seasoned Toppings and Sweet Toppings with recipes and suggestions for various uses.

Seasoned Toppings

These seasoned-just-right toppings will turn a basic casserole, steak, chicken, fish or salad into a company dish.

Seasoned Topping Number 1

1 c. ground gluten
2 tbsp. butter or margarine
1 c. Parmesan cheese
1/4 tsp. onion salt
1/4 tsp. garlic salt

Mix ingredients thoroughly. Spread thin on lightly-greased cookie sheet. Bake at 350 degrees for 12 to 15 minutes, stirring once. Adds extra flavor when sprinkled on spaghetti sauce, pizza, casseroles, tossed green salad, stuffed eggs or mixed with sour cream used as dressing for baked potatoes. May be used immediately or stored in closed container for future use.

Seasoned Topping Number 2

2 c. ground gluten
4 tbsp. butter or margarine
1 tsp. salt
1/2 tsp. basil
1/2 tsp. celery salt
1/2 tsp. pepper
1 tsp. paprika
1/2 tsp. brown sugar
1 tsp. dry mustard
$1 1/2$ tsp. Accent
1/4 tsp. ginger
1/4 tsp. sage or thyme
1/4 tsp. marjoram

Mix ingredients thoroughly. Spread thin on lightly-greased cookie sheet. Bake at 350 degrees for 8 to 10 minutes, stirring once. After it is baked, put it through the blender to make fine. Use as coating for chicken, steaks or chops by dipping each piece in beaten egg or buttermilk, then in topping and bake. One and one-half cups will coat a 3 pound fryer-size chicken. Store leftover topping in refrigerator.

Seasoned Topping Number 3

1 c. ground gluten
2 tbsp. butter or margarine
1/2 tsp. dry mustard
1 tbsp. Parmesan cheese
1/2 tsp. dill seed
1/4 tsp. nutmeg
1/4 tsp. garlic salt

Mix ingredients thoroughly. Spread thin on lightly-greased cookie sheet. Bake at 350 degrees for 12 to 15 minutes, stirring once. After it is baked, combine with sour cream or cream cheese and dry onion soup to taste for a chewy dip with chips or raw vegetables or mix equal parts with tuna and moisten with salad dressing for sandwich. Put through blender to make fine for use as coating for fish patties.

Recipes For Using Seasoned Toppings

Tuna Salad

1/4 c. *Seasoned Topping No. 3*
2 c. lettuce, chopped
1 can tuna, flaked
1/2 c. walnuts, chopped
2 boiled eggs, chopped
1/4 c. sweet pickles, chopped
1/2 c. celery, chopped
Salt and pepper

In large bowl combine all ingredients. Moisten with salad dressing. Chill. Serve on watercress.

Mayonnaise Dressing

1 tbsp. *Seasoned Topping No. 3*
1/4 c. mayonnaise
1/4 c. sour cream or IMO
2 tsp. onion, minced
Dash pepper, freshly ground

Mix together all ingredients. Serve on tossed salad. Has a chewy texture.

Sea-Food Dressing

1 c. mayonnaise
1 tbsp. Seasoned Topping No. 3
1/2 c. tomato, diced
1/2 c. cucumber, diced
1 tsp. onion, diced
Salt

Serve over fish patties and salmon loaf.

Creamed Eggs

4 hard-cooked eggs, quartered
1 can cream of mushroom soup
1/2 c. seasoned topping of your choice

Heat mushroom soup to simmering. Add topping and eggs. Simmer 2 minutes. Serve over hot biscuits or fluffy boiled rice or buttered whole wheat toast.

Stuffed Eggs

6 eggs, hard boiled
1/2 tsp. salt
1/4 tsp. pepper
3 tbsp. mayonnaise
1 tbsp. prepared mustard
2 tbsp. seasoned topping of your choice

Cut eggs in halves. Place yolks in bowl and add salt, pepper, mustard and mayonnaise. Mash with fork. Refill whites with yolk mixture and sprinkle with seasoned topping.

Dip For Raw Vegetables

1/4 c. *Seasoned Topping No. 1*
1/2 c. salad dressing or mayonnaise
1 c. sour cream or substitute
1 tbsp. sugar
1 tbsp. onion, minced
1/4 c. radishes, minced
1/4 c. cucumber, minced
1/4 c. green pepper, minced
1 tsp. salt
1 clove garlic, minced
Dash pepper

Combine all ingredients and mix well. Chill. Dip with raw cauliflower, carrots, turnips, zucchini squash, miniature tomatoes and green pepper. *Handy Crackers* from Using Bran and Starch Water section, potato and taco chips are also good. Makes about 2½ cups.

Casserole Toppings

Use 1/4-1/2 c. Seasoned Topping of your choice. Sprinkle over any casserole requiring topping. Bake as usual. Adds extra flavor, nutrition and good texture.

Blender Thousand Island Dressing

1 med. egg	2 c. any vegetable oil
1/2 tsp. prepared mustard	1½ c. catsup
	1 tbsp. onion, grated
2 tbsp. sugar	Dash paprika
2 tbsp. vinegar	1/4 c. *Seasoned*
4 tbsp. sweet pickle relish	*Topping No. 1*

Blend egg. Add next 4 ingredients and blend. Add remaining ingredients except topping and blend well. Pour in quart container and stir in topping. Makes about 1 quart.

Quality Blue Cheese Buttermilk Dressing

1/4 c. boiling water	1/2 tsp. onion salt
4 oz. blue cheese	1/4 tsp. salt
2 c. mayonnaise	1/2 tsp. Worcestershire sauce
1 c. buttermilk	1/8 tsp. hot pepper sauce (optional)
1/2 tsp. garlic salt	1/4 c. Seasoned Topping No. 1

Let cheese set in boiling water about 1 minute. Cream well. Mix in remaining ingredients. Makes about 1 quart.

Sweet Toppings

These helpful brand-new sweet toppings are a delightful shortcut to more nutritious desserts. They are delicious served as topping on ice cream, puddings, brownies, cup cakes, salads and desserts and as part nuts in cookies. Use them also in a Double Quick Pie Crust recipe and in ready-to-go cereal. These toppings store well in a closed container on the spice shelf if they are dried out.

Cocoa Topping

2 c. ground gluten
1/4 c. soft butter or margarine
1/2 c. brown sugar
2 tbsp. cocoa
1 tsp. vanilla

Mix together all ingredients, making sure butter is thoroughly mixed in. Spread thin on lightly-greased baking sheet. Bake at 350 degrees for 8 to 10 minutes. Stir 2 times. Do not get too brown. Use in pie crust for chocolate pie or fill crust with mint ice cream. Makes about 2 cups.

Cinnamon Topping

2 c. ground gluten
1/4 c. soft butter or margarine
1/2 c. brown sugar
2 tsp. cinnamon
1/2 tsp. cloves
1/2 tsp. vanilla

Mix and bake same as for Cocoa Topping. Use to make baked pie crust for fruit pies. Makes about 2 cups.

Brown Sugar Topping

2 c. ground gluten
1 c. coconut
1/2 c. brown sugar
1 tsp. vanilla
1/4 c. soft butter or margarine

Mix and bake same as for Cocoa Topping. Makes pie crust to be used as a graham cracker crust. Makes about 3 1/2 cups.

Recipes for Using Sweet Toppings

Double Quick Pie Crust

First blend topping to make fine. Sprinkle 1 to 1 1/2 cups topping of your choice over a well-buttered 8 or 9 inch pie pan. Bake at 325 degrees for 7 to 8 minutes. Cool.

Frozen Pie

3 c. milk	1 c. sugar
2 10 oz. pkgs. frozen strawberries or 2 c. fresh sweetened strawberries	2 c. cream, whipped

Mix all ingredients together except the cream. Put in ice trays and freeze. Take trays out, cut mixture in chunks and whip with cream. When creamy and smooth, pour in 2 *Double Quick Baked Pie Crust*. Set in freezer 20 to 30 minutes or until ready to serve. Makes 2 9" pies.

Fruit Pie

2 cans pie mix of either cherries, peaches or apples	1½ c. cinnamon topping 30 lg. marshmallows 1 c. milk
1 c. heavy cream, whipped	

Melt marshmallows in milk over low heat. Cool, then fold into cream. Butter generously 9 x 13 pan and sprinkle with 1 cup topping for crust. Pour fruit pie mix over topping. Spread evenly with marshmallow mixture. Sprinkle with remaining 1/2 cup topping. Chill 3 to 4 hours or overnight. Serves about 20.

Simple Ice Cream Pie

1 half gallon vanilla ice cream	2/3 of a 6 oz. can frozen pink lemonade

Let ice cream thaw half way, then whip in lemonade. Pour in 2 baked *Double Quick Pie Crust* and freeze till firm. Frozen orange juice or chocolate syrup may be used. Serves 12-16.

Frozen Fruit Pie

1-10 oz. pkg. frozen fruit either strawberries,	1/2 c. sugar, or more 1 egg white

raspberries, blueberries 1/2 c. heavy cream, whipped
or peaches
2 tsp. lemon juice

Mix fruit, lemon juice, sugar and egg white. Beat at high speed until mounds appear when beaters are raised. Fold into cream and carefully spoon into 2 baked Double Quick Pie Crust. Freeze several hours or overnight. Makes 2 9" pies.

Pineapple Dessert

1 lb. marshmallows
1 c. milk
1 lg. can crushed
　pineapple, drained
1 pt. whipping cream
2 c. topping, brown sugar

Place marshmallows and milk in double boiler and heat until marshmallows are melted. Cool. Add crushed pineapple and whipped cream. Sprinkle 1 cup topping in bottom of buttered 9 x 12 pan. Pour in pineapple mixture. Sprinkle 1 cup topping on top and lightly press in. Chill and serve. Serves 18-20.

Slice Date Cookies

1/2 c. brown sugar topping
1/2 c. nuts, chopped
1/2 lb. dates, chopped
15 marshmallows, small
30 graham crackers, crushed
Cream

Combine ingredients with just enough cream to moisten and stick together. Form into long 2 inch rolls. Set in refrigerator overnight. Slice to serve. Keeps well in refrigerator. About 2 doz. slices.

Special Fruit Salad

3/4 c. sugar
1 3-oz. pkg. lemon Jell-o
1 c. crushed pineapple
　with juice
1 c. hot water
2 bananas
1/4 c. cherries (optional)
1/4 c. sweet topping,
　brown sugar
1 c. small marshmallows
1/4 c. nuts, chopped
1 c. cream, whipped

Boil sugar and pineapple 5 minutes. Dissolve Jell-o in hot water and mix with pineapple mixture. Let jell half-way. Add fruit, topping, marshmallows and nuts. Fold in whipped cream. Chill until set. Serves 10.

Cream Cheese Cake

1 3 oz. pkg. lemon Jell-o
1 c. water, boiling
3 tbsp. lemon juice
1 8 oz. pkg. cream cheese
1 c. sugar
1 tsp. vanilla
1 can evaporated milk, chilled
2 9" *Double Quick Pie Crust*

Dissolve Jell-o in boiling water, add lemon juice and let set half way. Cream cheese, sugar and vanilla. Whip milk to stiff and fold in cheese and Jell-o mixture. Pour into pie crust. If desired, sprinkle with same topping used in crust. Chill and serve. 2 9" pies.

Marshmallows

2 c. sugar
3 envelopes unflavored gelatin
1/8 tsp. salt
1 tsp. vanilla
Powdered sugar
1 c. water
Sweet Topping

Combine sugar, gelatin and salt. Stir in water. Heat until sugar is dissolved, then bring just to a boil. Remove from heat, cool slightly, about 5 minutes. Stir in flavoring. Transfer to large mixer bowl. Beat at high speed of electric mixer about ten minutes or until mixture resembles marshmallow cream. Pour into buttered pan 13 x 9 x 2. Cool. Using kitchen shears dipped in hot water, cut into squares. Roll in favorite topping that has been put through blender. Makes about 6 dozen.

Refreshing Peppermint Dream

1 c. topping, brown sugar
2 c. marshmallows, miniature
1/2 pint whipping cream
2/3 c. crushed peppermint candy

Whip cream and combine with rest of ingredients. Chill and serve. Serves 6-8.

Ready-To-Go Cereal

4 c. rolled oats
3 c. rolled wheat
1 c. wheat germ
1/2 c. sunflower seed or nuts
1 c. coconut
1 c. sweet topping, your choice

1/4 c. honey
1/2 c. brown sugar
1/2 c. cooking oil
1/2 c. water
1 tsp. vanilla
1 tsp. salt

Mix together first 6 ingredients and set aside. In blender mix remaining ingredients, then add all ingredients together. Spread on 2 cookie sheets. Bake at 225 degrees for 2 hours. Stir occasionally. Cool, store in tightly-covered containers. Serve as breakfast cereal. Makes about 3 quarts.

Marshmallow Squares

1 lb. marshmallows
2/3 c. butter or margarine

1/2 c. topping, brown sugar
10 c. Rice Crispies, Cheerios or Trix

Over low heat melt marshmallows and butter in large sauce pan. Stir in remaining ingredients and pack in buttered 9 x 13 pan. Cool. Cut in squares. Makes 2 doz. squares.

Superb Cookies or Pie Crust

1/4 c. sweet topping, cocoa or brown sugar
1 egg
1 c. margarine

1 tsp. vanilla
1/4 tsp. almond extract
1 c. powdered sugar, sifted
$2\frac{1}{4}$ c. flour, sifted

Combine all ingredients to form soft dough. Form into a long $1\frac{1}{2}$ inch roll. Wrap in wax paper and chill. Slice 1/4 inch slices. *For cookies:* place on greased cookie sheet and bake 400 degrees 8 to 10 minutes. *For pie crust:* slice 1/8 inch slices and place around sides and bottom of a buttered pie pan. Bake 350 degrees about 15 minutes or until slightly brown. 4-5 doz. cookies. 2—9" pie shells.

Creamy Vanilla Ice Cream

6 eggs, separated
1 pt. whipping cream

1/2 tsp. salt
1 tbsp. pure vanilla

1 lg. can evaporated milk	Whole milk
2½ c. sugar	Sweet Topping

Separate eggs. Beat whites stiff. Gradually add sugar, beating until stiff. Carefully fold into whipped cream. Set aside. Beat egg yolks, add salt, evaporated milk and vanilla. Carefully fold into egg whites and cream mixture. Pour into 1 gal. electric or hand freezer. Fill to 1½ inches of top with whole milk. Freeze. Serve topped with gluten topping of your choice. Suggestion: Fruit may be added to ice cream mix. If fruit is sweetened use less sugar.

Family Home Evening Treat

1/2 c. Brown Sugar Topping	3/4 c. honey
1½ c. Powdered Milk (not instant)	3/4 c. crunchy or plain peanut butter

Mix together all ingredients. Form 2 or 3 long 1½ inch rolls. Roll in powdered sugar. Chill and slice in 1 inch slices. Suggestion: Omit topping and add 1/2 cup sprouted wheat.